D1648883

Climate Change:
An Archaeological Study

Climate Change:
An Archaeological Study

How Our Prehistoric Ancestors Responded to Global Warming

John D Grainger

PEN & SWORD HISTORY

First published in Great Britain in 2020 by
Pen & Sword History
An imprint of
Pen & Sword Books Ltd
Yorkshire – Philadelphia

Copyright © John D Grainger 2020

ISBN 978 1 52678 654 8

The right of John D Grainger to be identified as Author
of this work has been asserted by him in accordance with
the Copyright, Designs and Patents Act 1988.

A CIP catalogue record for this book is
available from the British Library.

All rights reserved. No part of this book may be reproduced or
transmitted in any form or by any means, electronic or mechanical
including photocopying, recording or by any information storage and
retrieval system, without permission from the Publisher in writing.

Typeset by Mac Style
Printed and bound in the UK by TJ Books Ltd,
Padstow, Cornwall.

Pen & Sword Books Limited incorporates the imprints of Atlas,
Archaeology, Aviation, Discovery, Family History, Fiction, History,
Maritime, Military, Military Classics, Politics, Select, Transport,
True Crime, Air World, Frontline Publishing, Leo Cooper, Remember
When, Seaforth Publishing, The Praetorian Press, Wharncliffe
Local History, Wharncliffe Transport, Wharncliffe True Crime
and White Owl.

For a complete list of Pen & Sword titles please contact

PEN & SWORD BOOKS LIMITED
47 Church Street, Barnsley, South Yorkshire, S70 2AS, England
E-mail: enquiries@pen-and-sword.co.uk
Website: www.pen-and-sword.co.uk

Or

PEN AND SWORD BOOKS
1950 Lawrence Rd, Havertown, PA 19083, USA
E-mail: Uspen-and-sword@casematepublishers.com
Website: www.penandswordbooks.com

Contents

List of Illustrations   vi

Introduction   1

Prologue: The First Australians   9

Chapter 1   The Ice Age   31

Chapter 2   Following the Ice   57

Chapter 3   Escaping the Ice   99

Chapter 4   Floods and Fish   134

Chapter 5   Sedentary Foraging   166

Chapter 6   Drought   201

Chapter 7   Agriculture   227

Conclusion   267

Index   285

List of Illustrations

Stages of the Ice Age   11
The Spread of *Homo sapiens*   13
Sundaland and Sahul   16
Early Sites in Australia and New Guinea   18
East New Guinea and Melanesia   21

GALLERY I: THE EARLIEST AUSTRALIANS   26

Effects of the Laki Eruption, 1783   35
Changes in Sea Level   40
Changes in Climate and Vegetation   41
Northern Hemisphere Vegetation Zones   42
Europe in the Ice Age   45
Caves in use in Ice Age Europe   46
The shrinking and division of Sahul   49
North America in the Ice Age   50

GALLERY II: EUROPE IN THE ICE AGE   52

Mammoth Hunters' Tent   63
Meiendorf and Stellmoor   65
Doggerland   68
Russia and Ukraine under the ice   72
Colonisation of Norway   78
Early Norwegian Ships   79
Arboreal Colonisation of Europe   82
The Retreat of the Ice   83
Beringia   89
Aleut and Inuit   92

GALLERY III: SCENES OF MESOLITHIC LIFE IN IBERIA 96

Ice and Water in Glacial North America 102
The Shifting Lake Agassiz 103
Lake Missoula 104
The Ice-free Corridor 112
Nenana and Denali Sites in Alaska 115
Early Sites in North America 118
Early Sites in South America 119
Early Sites in Alaska and British Columbia 120

GALLERY IV: EXTINCT ANIMALS OF NORTH AMERICA 131

Eastern North America under the Ice 135
Japan becomes an island 136
Mesolithic Shell Mounds 137
Early fishing in the Gulf 140
Alexander and Nearchos in the Desert 146
The Bass Strait Islands 151
Danish Mesolithic fishermen 155
A Danish Mesolithic House 160
A Canoe Burial 162
Japanese Pottery Decoration 171
The Periods of Japanese Prehistory 172
Japanese Shellmound village 175
The Foragers of Oronsay 179
Mesolithic India 188
Mesolithic Sites in the Gangetic Plain 189
Two Peruvian Early Sites 192

GALLERY V: MESOLITHIC LIFE IN INDIA 197

Changes in the Australian North 203
Sites in the Kakadu 206
Climate Changes in the Sahara and East Africa 212
Saharan Climate Fluctuations 213
Cattle in the Sahara 217

GALLERY VI: SAHARAN ROCK DRAWINGS   223

The Fertile Crescent   230
Ohalo   235
The First Agriculturalists   239
Natufian Houses   241
Early Agriculture in East Asia   248
Rice at Diaotonghuan Cave   250
African Domestication   255
East New Guinea   257
Kuk   260

GALLERY VII: THE INVENTION OF AGRICULTURE   265

Introduction

Global warming, we are told, is a fact. Evidence is produced almost daily to support this theory, large numbers of scientists are corralled into stating that it is happening, the United Nations has weighed in with conferences and reports that it is happening, large numbers of non-governmental agencies are involved – creating plenty of jobs for bureaucrats – lobbyists and agitators and non-governmental 'agencies' abound, seeking to exert pressure on governments to 'do something', and 'go green'. And yet it is actually not much more, so far, than a preliminary theory supplemented by projections of future trends, which vary widely with the hysteria of the projector. For the global temperature is still within the range over which it has oscillated for the last ten millennia. The problem is, of course, that the proof of the theory may well be an uninhabitable earth.

At the same time there are many who have disbelieved the prediction from the first. Some of these people have credible reasons for their scepticism. For example, one of the apparently most telling images which 'proves' that the warming is taking place is the reduction in the ice cover of the Arctic Ocean, which is popularly associated with the predicted rise in the sea level – but the Arctic ice is frozen sea water, and its melting will not seriously affect the sea level. At times the enthusiasts for global warming are guilty of allowing such 'facts' to go by without correction, imagining that such falsehoods assist their case. The ice caps and glaciers on land, however, are a different matter, for their melting will certainly cause a rise in the sea level. It would help, in considering the issues, if those who did so took a rather longer view than 'since records began', which is a disingenuous way of implying a long time. In fact, 'records' of climate rarely cover a period longer than the last two centuries; in climatological and geological terms this is the blink of an eye.

We will have to wait and see what actually happens, or rather our descendants will see, especially if we simply wait. Plenty of efforts are going into plans to restrict the warming, even reverse it, largely based on the theory that it is caused by the emission of carbon dioxide and other 'greenhouse gases'. This assumption may or may not be correct, and the suggested actions may or may not work. Again there are enthusiasts for 'solutions' which if implemented may well be so drastic as to be worse than the problem. It seems unlikely that we know enough about the Earth, its seas, and its air circulation, to risk fiddling with it any more. And yet the theory of global warming would not be the first scientific theory that proved to be mistaken, but which had the desired corrective effect.

It is worth remembering that diseases were regarded as Acts of God for many centuries, and that when malaria became a scourge it was ascribed to something in the air – 'mal-aria' = 'bad air' – which, in a (temporarily) convincing scientific theory of a 'miasma', indicated that stinking refuse and rotting waste should be removed so as to remove the miasma exuding from them, and when this was done, they thereupon ceased to produce the 'bad air'. Later came the germ theory of disease, and this debunked the miasma theory. But the dismissed and disproved theory had, all the same, indicated the right preventative action; removing the stinking refuse really did help to remove a major cause of malaria, at least in the temperate regions of the world.

Therefore, even if global warming really is happening, and even if the over-production of carbon dioxide and other greenhouse gases does not prove to have been the 'cause', reducing or eliminating the production of these chemicals can only be a good thing, and can only be a worthwhile exercise.

In fact, it is probable that the 'cause' of the warming is as likely to be the result of changes in the Earth's orbit and inclination, and in a change in the radiation received from the sun, as in man's production of gases. There is, therefore, much more to the issue than gases, for the Earth's air – a gas, note – is unstable. One only has to look out of the window at the moving clouds. The most important cause, however, is nothing to do with the atmosphere; it is all to do with humanity.

It is usual to call attention to the fact that the present episode of 'global warming' began once the Industrial Revolution had got under way in Western Europe and North America – since there is a heavy concentration on the idea that the increase in such gases as carbon dioxide is a main cause, therefore the burning of 'fossil fuels' must be the cause of the increase in atmospheric carbon dioxide and thus the cause of global warming. But that Industrial Revolution is also a period, the last two or three centuries, when the human population has been growing at a massive and unprecedented rate. This is a more likely basic reason for the change in the atmosphere, since it is that larger population which demands the products of industrial society and industrialised agriculture.

The ultimate background of all the agitation, theorising, talking, and effort over global warming is the fact that we are living on a planet which is subject to constant dramatic atmospheric fluctuations, and always has been. Which means that we, the human species, have been through periods of global warming and freezing more than once in the past, and, if we survive the present changes, these will happen in the future as well. On a minor key, there have been at least two 'little Ice Ages' in the last millennium, and the European Bronze Age was a warmer time than the present; in all those changes, the rise or fall in the temperature was not more than has happened in the past two centuries. That is, we have been here before, and did not panic, because the cause of the increase in the temperature was not known, but was surely enjoyed; it was a case of 'ignorance is bliss'.

The subject of this book, therefore, is the last time the earth went through a really decisive condition of global warming, and for this we need to go back to the end of the Ice Age. This took place between about 15,000 to 8,000 years ago, when such changes as melting ice, rises in sea level, the flooding of low-lying land, and a general increase in the temperature, happened. And all without the benefit of man's production of extra carbon dioxide or greenhouse gases, and with a minimal world population – that is to say, global warming on a massive scale has happened in the past, and obviously it was a natural phenomenon without any human agency involved.

Archaeology and geology are the sciences which revealed that man has faced the problem of this great climatic changes already. The great cold time came to an end with the worldwide phenomenon of the warming of the Earth by a few degrees, increasing the average annual temperature in Europe by about 7°C, which was enough to push the general climate from cold to warm. (An annual average reduction of such magnitude means a long period of winter where the temperature is well below that figure.) The men and women who lived during the Ice Age and while these subsequent changes were happening had to find ways of coping with the new and probably unwelcome situation. This is where archaeology comes in.

That the Ice Age people survived and coped with the new conditions is demonstrated by the fact that we, their descendants, are here. That they solved the problems they faced as a result of the warming, and flourished, is shown by the vast increase in the human population of the earth during the last 10,000 years. And this, of course, is the basic cause of the problem we face now (for increases in carbon emissions are primarily the result of increases in the human population). The problems which the Ice Age people faced were not in fact really issues of higher temperature, which many of them would probably have welcomed in itself; their difficulties were caused by the symptoms and effects of the change: the melting ice, the rising sea levels, the increase or decrease in local rainfall, any one of which phenomena might be the most important effect in a particular area. The methods by which the people of the Palaeolithic – the Old Stone Age, that is, also the Ice Age – adopted to cope with the troubles of their time of global warming turn out to be as various, unusual, and even unexpected as the symptoms themselves.

In effect, the various groups of people who lived through the changes conducted a series of experiments in different parts of the world – Europe, Japan, China, the Middle East, North and South America, Australia, the Sahara – in their efforts to cope with the phenomenon of the symptoms. Some of these experiments failed, some were briefly successful but led the people into dead ends, others were very successful. It was the results of these experiments, two of which turned out to be particularly successful in dealing with the problem, which has produced our own situation as it is today, with our wealth and our problems. That is, one of our problems in having to cope with a further episode of global warming is the result of our predecessors' success in coping with their own version of it. The methods which were used to help cope with the last period of global warming, in other words, have led on inexorably into the present condition of mankind, and had as one of its results, quite possibly, the present period of global warming.

I wish to present here, therefore, a survey of the several ways we have discovered by which people at the end of the last Ice Age, and even before then, attempted to deal with our present problem. It is only in the last century or so that it has been possible to do this; only in that time have all the investigations of archaeologists, ecologists, geologists, climatologists, glaciologists, and other scientists, revealed both the existence of the Ice Age and the life of mankind during and after it. One purpose of studying the past – there are many, of course – is to understand how people behaved in facing earlier crises and problems. The transition from the Ice Age was surely one of the greatest of these problems, as we now realise, and to see how those problems, which were seen by our ancestors in a variety of ways, were dealt with might yet provide us with some ideas, or at least an overall idea, for how we might cope with this new episode. I may also point out that the suggested warming at present is of the nature of one or two degrees; the people at the end of the Ice Age had to cope with a change of perhaps three or four times that.

In archaeological terminology, the period in question is called the Mesolithic, the Middle Stone Age. It is therefore a period which it is difficult to investigate, being so far in the past, with a relatively small population, and a time which lasted a few thousand years; in some accounts it is seen mainly as a time of transition between the Ice Age (the Palaeolithic) and the spread of farming (the Neolithic), both of which are much more interesting to many archaeologists.

And yet it will be seen that it was a time of much innovation, imagination, and achievement. For a small scattered population, existing in small communities which were largely isolated from each other, it is astounding that the people of the time advanced so much. It will be seen that, although they had little idea of how the new pressures on them had originated – though the people living close to the ice sheets must have understood that the temperature was rising – they based themselves firmly on their present condition in working out their solutions, elaborating the achievements of their own ancestors, and moving on from there. And the real surprise is that, despite the wide and isolated distribution of the small communities, many of them came up with the same or similar solutions to the world-wide problem, so much so that the issue of whether they were in communication with each other raises itself repeatedly. The answer to this has to be that they were not, which makes it all the more impressive that they were all procuring similar solutions and achieving much the same results.

It may also be argued that the global conditions and human capabilities have changed so much that comparisons between the Palaeolithic/Mesolithic and the 21st century AD are unrealistic. It is certainly true that economic life has changed, though broad economic principles are as fully applicable to the Ice Age as to today. But global conditions today are only marginally different in climatic terms, and human capabilities are very similar now to what they were in the Ice Age and after: the differences essentially are technological, and a consideration of life in the Ice Age shows that people in that time were as technologically alert and innovative as they are now; of course, they did not have television and nuclear weapons or the Internet, but making bows and arrows and spears, and manufacturing tools out of stone which are fashioned precisely for the purpose, shows that same mind-set as today (as Arthur C. Clarke suggested in his story 'Sentinel', the basis for Stanley Kubrick's film *2001*). It will emerge, therefore, that the basis of both times is the ability of humans to face change, cope with a major problem, and use the solution to their advantage. Certainly the climatic, technological, and intellectual conditions are different, but we are still the same species as the people who faced the drastic changes at the end of the Ice Age.

The purpose in this book, therefore, is to indicate the sort of options which were exploited in the past as an indication of the possible solutions which might exist for the future. This will not be an exact fit, in the sense that it is possible to look back and pick out a successful method of coping used in the past and then apply it today. This is partly because the successful methods of the past are in large part the ultimate cause of the present problem, but also because the past never repeats itself with any precision. On the other hand, there may be similar problems to be dealt with, and the fact that people in different parts of the world came up with similar solutions without communicating with each other does suggest that a collective answer will be possible.

And indeed, one difference is clear already. Where people at the end of the Ice Age were coping with the symptoms of global warming, today the aim seems to be to prevent the warming happening at all, an exercise, that is, in attempting to control the whole global climatic system. Such an attempt is a 'miasma' solution, in that it tackles the symptoms from the wrong premise, while containing within itself a recipe for a disaster greater than that which it is supposed to solve, and it evades the real issue, as does the attempt to regulate, or reduce, or abolish, the production of greenhouse gases. Such 'solutions' are short-term palliatives which will not work in the long term. The study of the Mesolithic populations' solutions to their similar problem may help to clarify present thoughts.

This survey may be thought particularly valuable in that the people who 'experimented' as the Ice Age ended did so in a general ignorance of what caused the changes they were faced with. One of the problems with the present situation, as I indicated with my reference to the search for the causes of malaria, is that we have begun by trying to detect the causes, and are going on from that theory – I repeat, it is only a theory, and may be mistaken – to advocate remedial action. In the same way, the people coping with the changes at the end of the Ice Age only dealt with symptoms and effects, and this is where we are now. In that sense the Mesolithic experimenters were the ideal laboratory animals, reacting in total ignorance of what was being done to them. In their reactions, successful or not, it should be possible to detect the requisite elements which will be required to cope with the new episode of global warming, should that be necessary.

Further Reading

Discussions of 'global warming' are already far too numerous to detail here, and are all too often ephemeral and inaccurate, overtaken or alarmist, and soon outdated. It is, however, worth mentioning, as a general account of recent climatic change, William J. Burroughs, *Climate Change in Prehistory, the End of the Reign of Chaos*, Cambridge 2005. One of the problems of the discussions on the issue is the involvement of scientists' egos; Burroughs seems admirably unaffected. A huge volume of descriptions of the world's climatic history was edited in the 1970s by H.H. Lamb, of which volume 2 is *Climatic History and the Future*, Princeton NJ 1977, which is especially useful and comprehensive. For a good general description, more archaeologically based, of the effects of climate on human life in the period following the Ice Age, see Brian Fagan, *The Long Summer, How Climate Changed Civilisation*, New York 2004.

Prologue

The First Australians

Before looking at the more immediate developments of the end of the Ice Age it is worth, as a preliminary, considering an earlier event, which occurred at the mid-point of that Ice Age. The 'Ice Age' is, of course, a vague but useful and compendious term for a long series of climatic events and changes which were spread over more than half a million years. In that time there were periods of low world temperature which alternated with warmer times (none of which changes, be it noted, were caused by human activities). The last phase, the Last Glacial Maximum as it is called, will be considered in more detail in the next chapter, for it seems that the decisive time so far as modern human beings were concerned was the rise in temperature which began at around 12,000 years ago. Such a period of warmer temperature is usually relatively brief, and the Ice Age earlier saw much longer warm periods between advances of the ice. Well before that, however, the region of Australia was the scene of a series of events which are also a part of the human response to an Ice Age, and which indicate some of the ingredients which will be seen again in the more recent time which is this book's main subject. (See Fig. 1, for the Ice Age fluctuations.)

Fluctuations in climate conditions have, of course, produced rises and falls of the ocean levels within the Ice Age. These changes are, of course, caused by the melting and freezing of ice caps on land, and result from alterations in the global temperature; the lower the temperature the more ice, and the lower the sea level. One of these periods of low temperature was particularly drastic and lengthy: from about 85,000 years ago the sea level fell more or less continuously for almost 20,000 years. The rate of fall bottomed out about 67,000 years ago; from then until about 61,000 years ago the temperature remained particularly low, and the world sea level was also therefore unusually low. Of course, the lowered sea level was the result of much moisture being held in the ice caps to the north and the south, in the Arctic and the Antarctic, and on the highest mountain ranges; this is the 'Heinrich 6' period in the 'Wurm' glaciation, in glacial terminology. As a result, the level of the sea was lower than at present by about seventy metres and that period of extremely low sea level lasted for almost 10,000 years.

Homo sapiens spreads from Africa

This period of glaciation was also the time when the new human species, whom we call *homo sapiens* – 'intelligent man' (a name which is a mark of our own egomaniacal species, and relegates other human species to a lower level, probably wrongly) – spread out from the original African homeland to the nearby lands, into southern Europe, Arabia, and India, and along the lands of Southeast Asia (Fig. 2). In this last area they were able to spread over a huge expanse of low land in Southeast Asia (which the glaciologists call 'Sundaland': Fig. 3). This is now partly flooded, but then it was dry, or at least not flooded. The peninsulas and islands of present-day Malaysia and Indonesia were then areas of high land. The wide plains stretched from west of modern Indonesia and enclosed all the Indonesian islands into one region of dry land which was then flooded.

Somewhere between 70,000 and 60,000 years ago, more or less in the last phase of the 'Heinrich 6' cold period, migratory people reached the eastern end of that territory, in what are now the islands from Lombok to Timor, and in Sulawesi. Beyond there they could see only the sea, with perhaps a few small islands in the distance.

These people were the descendants of an earlier people who had moved out of Africa tens of thousands of years before. They had an inherited tendency to migration, for their ancestors had moved regularly for generations into new lands, occupying and exploiting each territory by chasing and hunting eatable animals and moving about to gather eatable foods, fruits and vegetables and so on. The small societies of these earlier migrants tended to move to new sources of food, and they also repeatedly divided into smaller groups as they grew too numerous for the resources available. Some moved away from the primary family. Some turned southwards to move into southern Africa, but many appear to have gone northwards, and then eastwards and into southern Asia by

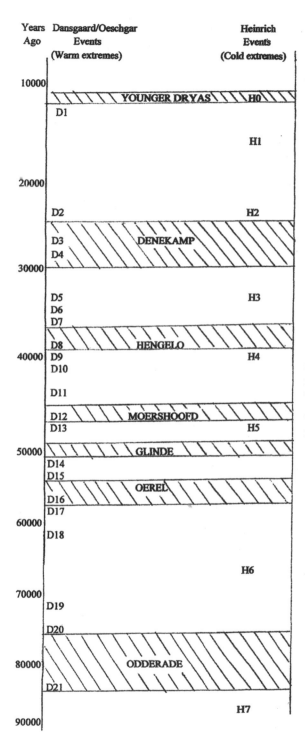

Years Ago	Dansgaard/Oeschgar Events (Warm extremes)	Heinrich Events (Cold extremes)

Fig. 1. **STAGES OF THE ICE AGE.** Ice Ages have happened for several hundreds of thousands of years, but it is only the last hundred millennia which are well studied and relevant to mankind. Here the alternation between cold ('interstadials') and warm periods is shown by shading, but the instability of the global climate is better indicated by the warm Dansgaard/ Oeschgar and cold Heinrich events marked on either side of the table. They clearly have happened irregularly so that warm periods may occur in interstadials and cold events in ice recessions. The last ten millennia, the Holocene, have been free of abrupt changes, other than fluctuations – that may well be changing.

way of Arabia. They were successful in their way of life, hunting animals, gathering fruits and nuts and vegetables and roots. In every generation, they multiplied, subdivided, and expanded further. On their migrations they encountered at least three other species of human beings, which have been given the names *homo neanderthalis*, 'Denisovans', and *homo florensis* (this last nicknamed 'hobbits') and yet another new species recently claimed for the Philippines. These last two groups were living in the very area our particular people had reached at the apparent end of their migrations, as they looked out at the forbidding expanse of sea south of Timor.

This expansion had been suitably slow at first – for the basic human population of *homo sapiens* was only a few thousand originally – but the movement speeded up about 80,000 years ago. It had taken perhaps 70,000 years for the migrants to reach Arabia and South Africa; it took perhaps only another 20,000 years for the descendants of the new 'Arabians' to reach Indonesia. And those who arrived at the eastern end of the Asian continent at the lands which later became Timor and Sulawesi clearly did not see any reason to stop simply because the land had run out and they could see nothing but sea. They set out to colonise Australia-New Guinea, a huge island continent, which was in fact out of sight from where they had reached, beyond the oceanic horizon.

People reach Sahul

A whole series of archaeological investigations during the last quarter of a century or so has now made it clear that representatives of *homo sapiens* reached Australia about 60,000 years ago or a little before. The same cold episode – 'Heinrich 6' – which had lowered the sea level to expose Sundaland did the same to the seas between Australia and New Guinea and between Australia and Tasmania; these three lands were therefore joined together by areas of low-lying dry land, forming a single continent which has been given the name 'Sahul' (Fig. 3). Australia itself was also considerably enlarged by the addition of the continental shelf and was linked to Tasmania by an area of low land which has now become the Bass Strait. The joint continent was perhaps a third larger in area than these several lands themselves are now. And yet there was always

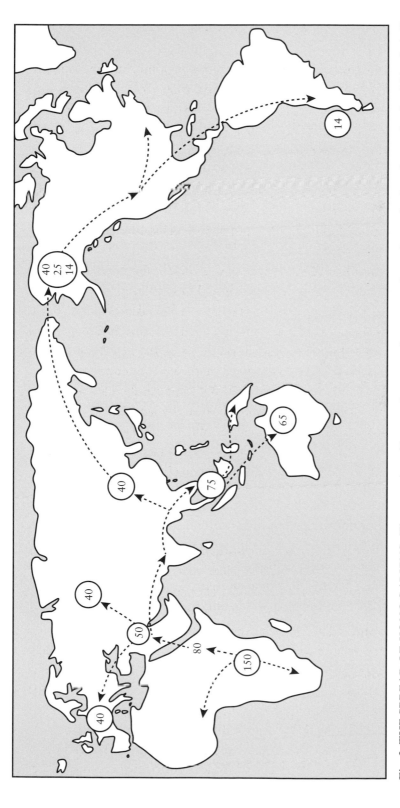

Fig. 2. THE SPREAD OF *HOMO SAPIENS*. The expansion of *homo sapiens* across the great continents began slowly, being confined to Africa for half of his existence. Once on the move, however, Australia was reached relatively quickly, and so the intervening lands as well. For the first 100,000 years he stayed in the tropics, only moving north into Europe and China in the long warm period between the Hengelo and Denekamp interstadia; there he was struck by the last long Ice Age. (Numbers indicate the approximate numbers of years ago areas were reached; arrows suggest the supposed routes. All numbers and routes are provisional and approximate.)

a seaway separating the continents of Asia-plus-Sundaland and Sahul, and, in particular, between an enlarged Java-plus-Sulawesi on the west and the continent of New Guinea-plus-Australia on the east. The sea was then, as now, dotted with islands – Timor, the Moluccas, the eastern Indonesian islands from Bali eastwards – which were all also larger than at present, and many of them were usually inter-visible.

In order to reach Timor the migrant people had already had to cross some narrow sea–straits, but in all cases the crossings were fairly narrow, and in particular the target lands could be seen all the time. But from Timor and Sulawesi onwards the situation was different. It was not possible to see more than some small islands from the shores of the Asia-Sundaland continent; Australia-Sahul was quite certainly well out of sight.

The Earliest Evidence

Enough instances of well-dated human remains and other evidence of human activity have now been found in various parts of Australia to make it certain that men, the ancestors of native Australians ('Aboriginals') reached Australia about 60,000 years or more ago. At least two sites, that at Lake Mungo in western New South Wales being the most important, have been dated to between 50,000 and 60,000 years ago, and another to 62,000 years ago. (All these early dates are very approximate, being the result of a combination of various methods of dating materials, such as radiocarbon (whose limit is about 40,000 years before the present) and thermo-luminescence, neither of which, at this time distance, can be more precise.) There are several sites in Tasmania dated to between 35,000 and 11,000 years ago, and two in Queensland which are rather older. Men had also reached New Guinea, where a cave on the Huon Peninsula at the eastern end of the island has produced a date of about 40,000 years ago, and other sites in the rainforest of the island have been dated to between 35,000, and 32,000 years ago (Fig. 4).

Some of these dates may be argued over, but collectively the whole suite of dates from eastern New Guinea to Tasmania is convincing. Their significance is that they show that the ancestors of the people who lived at these places had to have crossed from Sundaland to Sahul at some time before the dates of the earliest remains. How long will it have taken the descendants of the earliest arrivals in Australia – necessarily in the north – to get from the north coast to New South Wales, where they were already living by 60,000 years ago? Perhaps not very long. They were in a new continent, with new animals to hunt, and which had a large desert in its centre. They probably moved fairly quickly throughout the whole island. Their presence also, more importantly for the point of view of this book, must mean that the first arrivals had made a sea journey of at least several days, some part of which was spent out of sight of land, in order to reach their destination – and which they could only have discovered by preliminary exploration.

Sailing to Sahul

There were two possible routes by which people could have reached Sahul from Indonesia: from the north of Sulawesi by way of an enlarged Halmahera into the western part of New Guinea, and a southern route from Java by way of the string of (enlarged) islands as far as Timor and then south-east across the Timor Sea into Sahul. Neither of these can be said to have been either easy or straightforward. A cave site on the western tip of Timor was occupied 30,000 to 35,000 years ago: in New Guinea, the Huon Peninsula site on the east is even earlier; we can say, therefore, that both routes were probably being used, but the sea distances were different, as were the destinations. (The northern route required two voyages, both fairly lengthy; the southern needed many voyages, but all but one of them were fairly short.) The different routes also took the voyagers to differing destinations.

Those who moved due east from Sulawesi became the ancestors of the inhabitants of New Guinea, and, because of the densely forested interior of the island, they were probably largely confined to sites along the north coast for a long time; following the simplest routes of expansion, they moved along that coast, always eastwards, and eventually they came to another wide sea. Since they had arrived by sea and continued to live along the coast, it would be reasonable to assume that they retained the use of boats all this time.

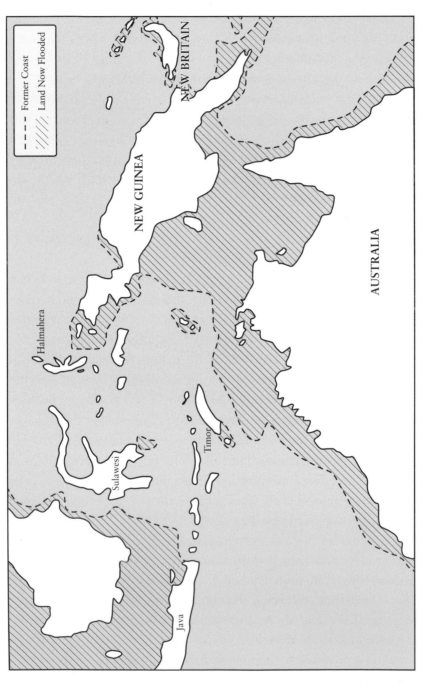

Fig. 3. SUNDALAND AND SAHUL. The sea which faced the colonisers of Sundaland (Australia/New Guinea/Tasmania) 60,000 years ago or so was formidable. Shaded areas are now flooded land. The colonisers appear to have set off from Timor, crossing the Timor Strait, and from Sulawesi towards Halmahera. To get to those islands they had already had to cross narrow seas – but the next step was to reach land which is out of sight of their starting points.

Those moving out of the area from Timor went southwards. Actually they could have sailed from Timor in any direction between south and southeast and they would have reached the new continent. Given the uncertainties of sea voyaging in their necessarily primitive craft, different groups very likely landed in a variety of places. But from the landing places, wherever they were – but clearly on the north-west coast of Australia – they had before them the whole continent of enlarged Australia. In many areas it was as dry as today (for the reduced sea level and the retention of moisture in the ice caps also reduced precipitation everywhere), but it was certainly hospitable to hunters and gatherers, at least once they had got used to the peculiar set of animals which had earlier evolved on that long-isolated island continent.

We have to imagine that the migrants sent prospectors ahead to investigate the possibilities – even to see if there was land in the first place – and that whole families later made the crossing. They voyaged in sufficient numbers to inject a viable human population into the new land. We must assume that several families at least made the crossing, a minimum perhaps of several dozens of people, men, women, children, old and young, though possibly more than that; it is also likely that the voyages were repeated by others at later times, bringing new populations to the new land. Boats of some size and reliability were essential and perhaps had to be invented for the occasion, though they had had to cross narrow straits to get as far as Timor. They had used boats to cross the straits between the islands, but the voyage from Timor to Australia was a much more difficult proposition and would probably have required bigger and more reliable craft. Probably the whole migration took place over a long period of time, for once the first migrants had proved that it could be done, others would probably follow.

It was clearly necessary to plan ahead with some care in arranging and accomplishing all this. It was not a matter of 'sailing to Australia' (or Sahul), and stepping ashore like immigrants arriving at Sydney or Melbourne and landing on the dockside. We have to assume that a series of crossings was made in each case, as the scouts prospected the new lands and then returned to their original homes, where much discussion will have taken place as to whether to risk the move. The several migrant families then crossed, and some had perhaps to abandon their first attempt

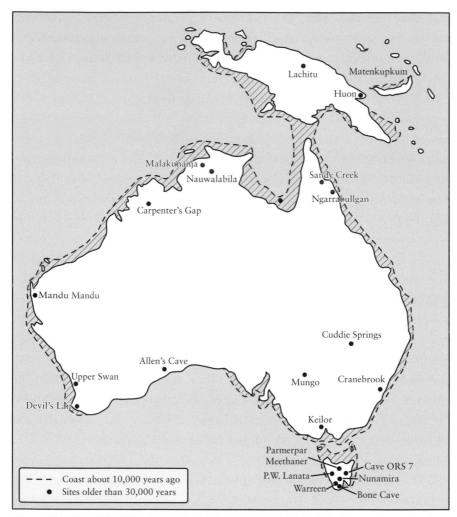

Fig. 4. EARLY SITES IN AUSTRALIA AND NEW GUINEA. Recent investigations have demonstrated the early settlement of Australia, New Guinea, and Tasmania. The sites shown on this map are all well dated to more than 30,000 years ago; two of them, Lake Mungo and Malakunanja, are twice that age. The discovery of these sites in part depends on the occurrence of rock shelters and caves where material remains have been preserved; it also depends on the enthusiasm and activity of researchers, which is in part why so many early sites are known in Tasmania.

and return and try again, so these communities probably accomplished these crossings several times in both directions. Undoubtedly there were casualties, families who were drowned or wrecked, or ran out of provisions in mid–ocean, and prospectors who simply vanished – and unlike the migrants trying to cross from Africa to Europe nowadays there were no rescue ships available. No doubt also, different voyages landed at different places; the several groups may well have lost touch with each other for some time.

'Proofs'

This has been a tailor-made situation for experimental archaeology, and sure enough an archaeologist has sailed from Timor to Australia on a raft which he imagined could have been built by the travellers of 64,000 years ago, using such natural materials which might have been available to Palaeolithic voyagers – bamboo, vegetable binding, and so on. His voyage took 12 days, though it was only five days from Timor to where the coast of Sahul would have begun. Professor Robert Bednarik, who organised the expedition, noted that some sort of rudimentary steering system was required, and he gave the craft a sail, which seems an unlikely invention at that time, so the craft he sailed was not really in any way 'authentic'. Above all, he knew where he was going, and that there was land awaiting him at the end of the voyage (not to mention a rescue craft undoubtedly standing by). And then, of course, all that he proved was that the crossing was possible, and we knew that already. But it helped to silence some of the sceptics, for there are always nay-sayers who claim that 'primitive' people could not possibly be inventive, brave, and clever. Once those early archaeological dates are accepted, however, the rest – the voyages, the explorations, the bravery, the planning, the human intelligence – automatically follows.

Occupation of Australia

These ancestors of the native Australians had a large continent to explore and exploit; they spread throughout the continent fairly quickly, in the same way as their ancestors had moved from Africa as far as East Asia (and, it seems, just as the early Americans occupied their own new double continent at speed (see Chapter 3)). The earliest dates of human habitation in Australia are from both the north and south of the continent. The very earliest are actually from the site at Lake Mungo in the south, already mentioned. From Malakunanja in the north, in Arnhem Land, three dates of between 60,000 and 50,000 years ago are thus a little later (and these dates are less controversial than are those from Mungo). This northern area was habitable, consisting of open forest and steppeland, and though the centre of the continent was as dry as it is today, that dry region was surrounded by wetter and habitable lands. Three varieties of the method of colonisation have been theorized – radiating out from the landing area, encircling the dry centre by moving along the coast, and a mixture of the two, not that the migrants were operating to any theory.

The interior of the continent may well have been exceptionally dry at the time of the earliest crossings, with a large central desert, but the rise in the sea level as a result of the period of global warming which began about 61,000 years ago, increased the inland moisture (and raised the sea level to flood the lower parts of Sahul so 'isolating' the descendants of the pioneers). Lake Mungo, for example, was one of a group of lakes well filled with water at the time the earliest human occupants were there. (See Gallery I.)

On to Melanesia

The migrants into Australia, with a huge continent before them, soon became landlubbers once more. Those who moved into New Guinea, however, did not abandon the sea. Faced with a thick tropical forest and high mountains inland to their south, they moved along the more accessible north coast, so the oldest site in New Guinea, at Fortification Point on the Huon Peninsula, was close to the coast even when there was a lower sea level than now. Given the forests which come down to the shore, this journey would most easily have been accomplished by sea.

The differences between the lands which the two sets of voyagers found dictated their different responses – with a wide open and forested land before them, those who reached Australia abandoned their boats and walked, hunting the animals – the strange animals – they found; in New Guinea, by contrast, it would be easier to keep using their boats in travelling along the coasts. From the Huon Peninsula and neighbourhood, the New Guineans kept on going eastwards, necessarily again by sea. There are early settlements on the islands of New Britain

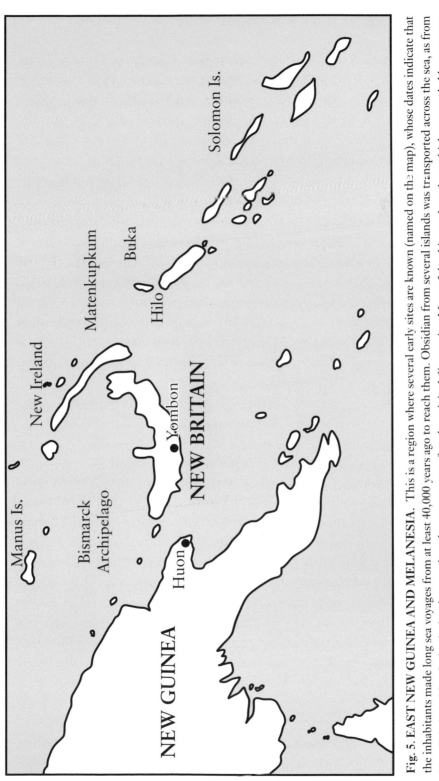

Fig. 5. EAST NEW GUINEA AND MELANESIA. This is a region where several early sites are known (named on the map), whose dates indicate that the inhabitants made long sea voyages from at least 40,000 years ago to reach them. Obsidian from several islands was transported across the sea, as from New Britain to New Ireland, showing continued sea transport after the original discoveries. Manus Island is another place which was settled by voyagers moving out of sight of land for some distance.

and New Ireland – sites are dated to 35,000 years ago at Makenkuptum on New Ireland, and at Yombon on New Britain, and another is of 28,000 years ago at Hilo on Buka Island, the northernmost of the Solomon Islands (Fig. 5). To reach these islands, the migrants must have travelled by sea, even when the sea level was lower than it is now; and to reach Manus Island they had again to travel out of sight of land, which implies preliminary blind exploration. It is best, therefore, to assume that this branch of the original population retained the use of boats from the time of their very first explorations, whereas the Australians did not need to – though Australians who later lived in various places on the coast usually took to the sea when it was advantageous.

This was, again, not simply a one-way movement into the Melanesian islands. On New Britain there are two places where obsidian can be obtained. This is a distinctive black and shiny volcanic glass which when split gives an exceptionally sharp and durable edge, as well as being shiny and of considerable beauty. Wherever it has been found it is so valued that it was normally widely exported, or at least transported to other places. It is therefore a particularly useful indication of human movement in the non–literate centuries, since each obsidian source produced a distinctive type of stone whose chemical make-up and internal structure enables it to be used to trace movements by the people who used and carried it; it was probably one of the earliest of trade goods. Obsidian from New Britain is found in early settlement contexts on New Ireland, having been moved several hundred kilometres. The human settlers of the islands of Melanesia were therefore probably in regular and continuous contact with each other by sea from the time they arrived – a period of at least 40,000 years, and probably, given the earliest Australian dates, half as much again. They were the first peoples anywhere to use the sea in this way – that is, as a regular and repeated means of communication, not just for a single crossing; eventually they contributed their skills to the peopling of the Pacific Islands by the Polynesians.

Hunters and Gatherers

The nature of the human society on these islands, as reflected in the archaeological findings, was much the same as elsewhere in the Palaeolithic world: moving bands of hunters and gatherers who did not settle in any one place for very long, so that the evidence for their presence is in the form of small campsites where remains of food, bones, stone tools, and sometimes human remains, may be found. These were sites which were occupied only briefly, perhaps for a few days or weeks or for a single season; some of them might be occupied repeatedly over a period of several years, with the hunters returning for a brief stay at a favourite spot, before being wholly abandoned.

It is remarkable enough that the traces of these very early settlements have survived and been identified, and at such immense time distances, but the story they tell makes it even more worth considering than just as evidence for the earliest occupation of the huge continent. A human skeleton found near Lake Mungo, and nicknamed 'Mungo Man', had been deliberately buried in a shallow grave dug into the sand dune overlooking the lake (which is now dry). His body had been sprinkled with red ochre powder which had been brought from some distance away and which was obviously manufactured from rock deliberately excavated and ground into powder. This is a ceremony, using the same material, which is known in other lands and societies; it is evidently a process designed to pretend that the deceased was leaving for a new life, and a reminder of his liveliness when alive. He was of a light build, referred to as 'gracile' (that is, slim and small and neat), which has led to suggestions that he was of a different human species than *homo sapiens*. The implications of the red ochre, and of the site and configuration of the burial, all suggest that the people who buried him believed that he would have a further life once he had died. His grave was deliberately placed to overlook a camp where he had presumably lived, which suggests an appreciation of his former life and fond memories of the place. All this is a burial method and context which can be paralleled in America, Asia, Europe, and Africa, though not as early as this example in Australia.

The other very early date for Australian settlement comes from the north, in Arnhem Land where stone tools dated to more than 50,000 years ago have been found in the rock shelter at Malakunanja. (Here there is no argument as to the integrity of the dating, whereas there are still some disputes over the Mungo dating.) It was a site occupied continuously (or perhaps, more likely, repeatedly) for perhaps 8,000 years, and then more intermittently for more thousands of years; it was evidently a place which was immediately attractive to hunter gatherers. There is some indication, in the form of red and yellow ochre once again, ground haematite and mica, that the earliest inhabitants decorated themselves with paint. At least one other early site in Australia, at the Carpenter's Gap rock shelter in the Kimberley area of Western Australia, has produced some evidence that the interior of the shelter was probably painted, or at least coloured, possibly 40,000 years ago; these people had a visual appreciation of colour and beauty, and the time to prepare for and indulge it.

The point of all this is to demonstrate that the habits of mind of the earliest people who inhabited Australia were similar to those who were still in Africa, of those in Asia, and of those who moved into the frozen wastes of Europe some considerable time after Australia was colonized – and to modern humans. *Homo sapiens* is seen to be inventive, exploratory, courageous, and artistic wherever he went, and to have had a belief in an afterlife, which argues a very vigorous imagination. The early inhabitants of Australia were relatively-recently related to the other groups who had moved into other parts of the Old World. (And there are some signs – widely disbelieved, of course – that some of them may have got into America as much as 40,000 years ago, but if the lower sea level joined Australia with New Guinea and Tasmania, it also joined Asia and America, and people would be able to cross from one to the other.) It is clear that human beings had by that time already become capable of facing a variety of climatic conditions and they were prospering in a wide variety of lands, including desert, coast, the sea, and forest.

These early Australians and Melanesians were therefore particular, and particularly early, cases of people coping with the results of the changes brought about by the Ice Age. In this case, of course, it was the result of the expansion of the ice caps and of a lowering of the sea level, but their movements are indicative of the adaptability of the species, and of the ability of people to deal with unusual circumstances. In northerly latitudes it was the more distant relatives of the first Papuans, Australians and Melanesians who were soon to move north and adapt themselves to the possibilities of life in the icebound lands of Europe and China – as different a climatic regime from that of Australia as can be imagined. It is worth remembering that these people originated in Africa, an area which was affected only marginally by the ice, yet their descendants took to the sea to reach Australia, developed a maritime society in Melanesia, and became competent big-game hunters and artists in ice-covered Europe. Adaptability and inventiveness are the major human characteristics.

Further Reading

The most accessible and comprehensive discussion of Australian prehistory is by Josephine Flood, *The Archaeology of Dreamtime, the Story of Prehistoric Australia and its People*, revised edition, Marleston, South Australia 2004 (which contains a 'Stop Press' on the latest discoveries, including a note on the Carpenter's Gap rock art); this may be supplemented by her earlier book, *The Riches of Ancient Australia, a Journey into Prehistory*, St Lucia, Queensland 1990, which is a guide to the visible remains.

For more technical reports see the supplement to *Antiquity*, vol. 69, 1995, called 'Transitions, Pleistocene to Holocene in Australia and Papua New Guinea', though more work has been done since then. Professor Bednarik's voyage is described in two articles in the *International Journal of Nautical Archaeology*: 'The Earliest Evidence for Ocean Navigation', and 'Nala Tasih 2: Journey of a Middle Palaeolithic Raft', 26, 1997, 183–193 and 28, 1999, 25–33. Early settlement in Melanesia was discussed by Jim Allen, Chris Gosden and J. Peter White, 'Human Pleistocene Adaptations in the Tropical Island Pacific: Recent Evidence from New Ireland, a Greater Australian Outlier,' *Antiquity*, 1980, 548–561.

The new human species from the Philippines was reported in *The Economist*, 13 April 2019, 76–77 and *Current World Archaeology*, 95, 2019, 4.

Gallery I

The Early Australians

Two locations of particular importance in the early history of Australia – the Willandra Lakes area in western New South Wales, and Kakadu in the Northern Territory. The fact that the Willandra area is so far south, and yet has produced archaelogical remains of a very great age – 60,000 years – shows that the country was very quickly occupied by its first invaders; the Kakadu is one of the few regions where aboriginal life continues to this day, the inhabitants being direct descendants of those early invaders, whose landfall was not far away.

I.1 Landscape in the Willandra Lake area – today.

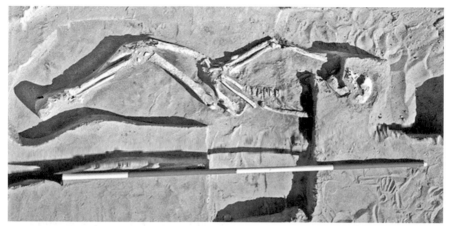

I.2 The Lake Mungo burial, a small man but buried with care and affection.

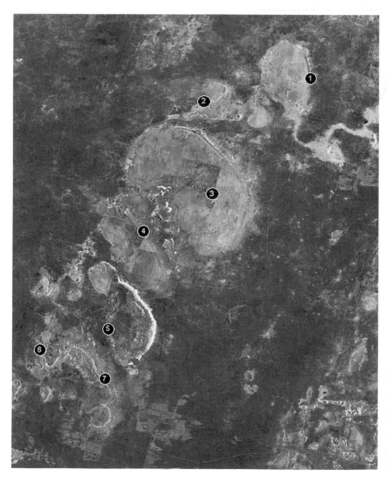

I.3 The Willandra 'Lakes' from the air – dry and dessicated today.

The earliest paintings date back, as far as can be ascertained, at least 30,000 years, though since the first Australians were using red ochre before then, it is likely that the earliest pictures were even older. An early version was hammered designs into rock (I.6), and the outline drawings in red are probably later (I.5). The tradition of painting on rock walls continued at least into the 1960s.

I.4 Kakadu landscape – hot and wet, forested and hilly.

I.5 The painting of a thyacine, an extinct 'marsupial wolf'.

I.6 An early hammered carving showing various native Australian activities.

I.7 A Kakadu rock shelter, with paintings.

Chapter 1

The Ice Age

There have been a long series of Ice Ages, or perhaps it would be best to say that the Ice Age consisted of a series of advances and retreats of the ice (Fig. 1). That is, the temperature of the surface of the earth, and of the air envelope around it, has always fluctuated. As the temperature fell, so the water vapour in the air condensed into ice, and lay upon the land and on the highest mountains and the lands in the world's north and south; this in turn meant that the oceans shrank as less water reached the land. Later, as the temperature rose, the ice melted, the water and the water vapour were released once more, and the oceans and the land were replenished. It was a natural process, dependent on radiation from the sun.

The temperature of the earth has in fact never been constant. Even during the last 12,000 years or so, when we, self-centred, simply assume that the Ice Age is now well in the past, there have been many minor fluctuations; the Later Middle Ages in Europe have been called the 'Little Ice Age', a term also used for the cold time in the eighteenth and nineteenth centuries; the Bronze Age was a period of warmer climate than at present. That is to say, we are still living in the climate of the Ice Age, which has been for 10,000 years or so in a temporarily almost-stable condition – described, again in a typically human self-centred way, as the 'climatic optimum'. It must also be said that the world's climate was also almost stable when ice covered much of the northern hemisphere; stability is, as always, a question of perspective.

This has in fact been the constant condition of the earth. Ice Ages have been detected up to 750 million years ago, in the Pre-Cambrian geological period, which is as far back as it is possible to examine the rock formations which have preserved the evidence of the activity of the ice. But it is only in the last 100,000 years that they have been important for human beings, since it is only in that period of time that *homo sapiens* has existed. (Earlier varieties of humans – *erectus, habilis, neanderthalis*, and so on – are not our concern, though it is obvious that they were similarly affected by advances and retreats of the ice in their times; the assumption that earlier *homo* species, such as *neanderthalis* and the barely-known Denisovans, died out in the last part of the Ice Age (that is, during the Last Glacial Maximum) may be true, but this is not necessarily effect and cause; both human species were well adapted to life in the cold, and had been for millennia.)

It is this last set of Ice Ages, over the last hundred millennia, which are the best known, for the evidence is much clearer in the rocks and the seas, and in the ice of Greenland and Antarctica. This was the age which has affected modern man, and in which man's abilities were honed and sharpened. From *homo sapiens'* perspective the Ice Ages may have been climatic catastrophes. Global warming and freezing has therefore been the condition of the world within which modern *homo sapiens* has evolved. This also means that the evolution of man into *homo sapiens* has been strongly affected by the Ice Ages and their subsequent amelioration and fluctuation. One might almost say that *homo sapiens'* evolution came about as a result of the Ice Age.

Causes

The causes of the fluctuations of the climate are complex, and a number of factors enter into the equation. Fluctuations in the radiation received from the sun is one part of it, as already mentioned – though in that case it simply puts the question one stage further back, since one then has to try to explain why the sun's radiation should fluctuate; that is not an issue I wish to enter into here. Alterations in the Earth's orbital pattern have at times reduced the reception of radiation in certain areas, such as the northernmost latitudes. In the very remote past changes in the distribution and location of land and sea had their effect, though such changes – continental drift, for example, and mountain building – have not been very prominent in the recent past and during man's career, because of their slow nature, and because of the relative brevity of *homo sapiens'* existence so far. At times major volcanic explosions can have serious effects on the quantity of solar radiation reaching the Earth, though this, by contrast, tends to be a fairly short-term matter, since the dust which the erupting volcanoes produce, and which is what affects the climate, dissipates relatively quickly. And, of course, in the last three or four centuries, man's production of the 'greenhouse gases' has been added to the mixture.

The significance of whatever theory or mixture of theories is accepted, of course, is that the process of climate change in the past was always a natural phenomenon; the present episode of global warming is, by contrast, claimed to be the 'fault' of human intervention, notably industrialisation and the consumption of fossil fuels. It is, however, sometimes difficult to distinguish the various components of the changes and their causes under the 'natural' regime – not helped, of course, by human passion and prejudice and argument and monomaniacal thinking, though that is a part of the human condition – and the causes of the present changes are equally various and complex. However, since in the past these changes have been entirely a natural phenomenon, then that must obviously be the starting point for research into the causes of the present changes, an obvious point which has been largely ignored in favour of simply blaming the problem on mankind's activities, a new version of the curious theory of original sin.

In fact, the actual cause or causes of the present changes is not relevant to a consideration of past episodes. Whatever particular causes were or are dominant at any time or period, the effects are likely to be much the same. Human responses may turn out to be similar in each episode, and that does apply to our present condition as much as to these events in the past. The change is certainly happening, for whatever cause or causes, but it is the effect on the earth and therefore on mankind and the other inhabitants of the earth which is at issue here.

For the past 500,000 years, the quantity of ice at the North and South Poles, and on the highest mountain ranges, has changed year by year, and it continues to do so. The change is usually relatively slow, though it has been possible to measure it by annual changes. Once the geologists began to investigate glaciation, in the early nineteenth century, the process became clear. The retreats and advances of the ice are phenomena which feed on themselves, and once either process is happening it tends to go on for several millennia. If the radiation reaching the Earth is reduced for one reason or another this causes a fall in the air temperature which causes precipitation to take place in the form of snow, rather than rain; the reduced air and ground temperature prevents the snow on the ground from melting; in turn, this snow cover reflects back more of the radiation of the sun, and so the air temperature is reduced still further, and the next set of precipitation upon the cold, snow-covered ground is again as snow. As the process continues the snow compacts under the weight of successive snowfalls and forms ice. The increasing weight of the ice forces out tongues of itself in lower land, particularly along valleys – hence the glaciers. The main effect, however, is in the wide covering of ice over huge continental areas; the glaciers are only symptoms of this, but their changes have been the most sensitive and obvious mark of climate change. The great ice sheets are the main forces for the stability of the colder climate for they reflect away the sun's radiation, thus preserving the cold. The reverse effect, global warming, similarly continues over a long time once it has begun; it does not take much of a change in the temperature to set the process going, but a change of only a few degrees, if prolonged and maintained, will have serious long-term effects. It is clear that we are in such a condition now.

The process is not, however, smooth and inevitable. Sudden events can trigger a catastrophic temperature change. One Ice Age ice-advance may have been set off, or at least pushed on, it seems, by the massive volcanic explosion of Toba in Sumatra about 74,000 years ago. Such events throw great quantities of dust into the atmosphere which then reflect back the sun's radiation – in effect casting the land beneath into shadow. Even much smaller volcanic events can have great effects. The eruption of Laki in Iceland in 1783 spread a dust cloud which reached over Canada, Alaska, Siberia, and over all Europe, as far as Baghdad. The result was acid rain, bad harvests, starvation, disease well beyond the regions directly affected by the dust cloud (Fig. 6). The subsequent famine lasted three or four years in some lands. Japan, on the far side of the world from the volcano, had famine for three years, the result of a lowered temperature.

The eruption of Toba in Sumatra 74,000 years ago left a caldera (the wrecked volcano crater) 100 kilometres long and 60 kilometres wide – now containing a lake – and threw an estimated 3,000 cubic kilometres

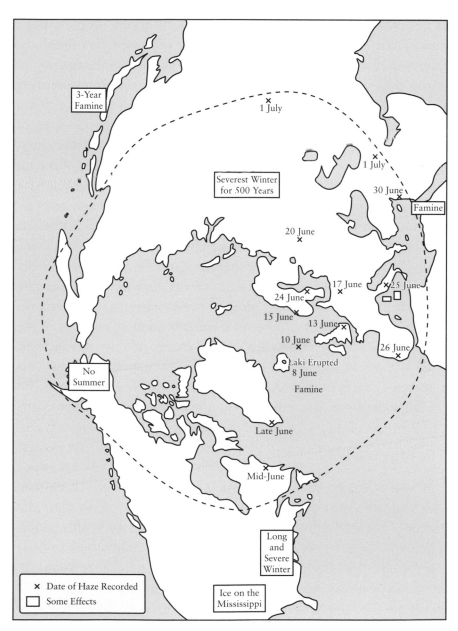

Fig. 6. EFFECTS OF THE LAKI ERUPTION, 1783. The volcano Laki in Iceland erupted in June 1783. The resulting cloud of ash and fumes covered much of the northern hemisphere, reducing global temperatures, and extending winters; later effects included famines in Egypt, Alaska, and Japan. Iceland suffered the worst of all, of course – a three-year famine, and a quarter of the population died. In England and France there were several tens of thousands of 'extra' deaths in 1783, largely because of the poisonous fumes.

of rock and dust into the atmosphere; up to 35 centimetres of ash was deposited in the Indian Ocean as far as 2,500 kilometres downwind; it is estimated that the dust thrown into the atmosphere caused a drop of 5°C in the global temperature, and three times that in the coldest northern latitudes. This event has been positively linked to a new and more intense phase in the Ice Age, though cooling had already been going on for several millennia before the eruption. It is this Ice Age advance which so reduced the sea level of the oceans that men were able to cross into Australia – and perhaps into America (see the Prologue).

This eruption is the sort of sudden event which, while generally short-term in its direct and immediate effects, might well initiate the process of wider change or intensify one which was already happening. Such an event, however, will only have an ongoing effect if that earlier, more gradual, change was already under way. Otherwise, like the Laki eruption which occurred in a period of global warming, the effects will relatively quickly dissipate and have no long-term significance. Thus, only if the sudden shock fits with what is already going on will the long-term changes in the climate be effectively permanent. Normally the climate will not be derailed even by a massive volcanic explosion; such an explosion, however, may well speed up the changes.

The Effects on Man

Ultimately, the process of freezing and ice accumulation becomes stabilised, with extensive areas covered by thick sheets of ice, but in the process, the physical and vegetational geography of the world changes. The quantity of ice deposited on the land was concentrated on the highest mountain regions and at the poles, from which it spread out to lower altitudes and latitudes, both in the form of growing ice sheets and as glaciers flowing along the valleys. The water locked up in this ice reduced the level of the ocean, so the areas which are now shallow parts of the oceans, that is, the continental shelfs, became dry land – as with Sahul and Sundaland (Fig. 3), the Sea of Japan (Fig. 33), the Persian Gulf (Fig. 35), the North Sea ('Doggerland') (Fig. 16), the Bering Strait (Fig. 22), and other areas. For mankind this meant that certain areas were effectively now out of reach because of the covering of snow and ice; yet at the same time other areas away from the icefields became available for habitation due to the lowering of the sea level. For a species which, during the whole of the Ice Age, essentially consisted of a series of communities of migrating hunters and gatherers, the quantitative difference was probably of little or no importance. What was lost to the ice was gained by the fall in the ocean level, and the animals and the hunters moved to the new areas. (Note that as the sea level will rise in our present episode of global warming, so will the frozen lands of the north become available: Canada and Alaska and Russia are already noticing changes.)

The effect of the ice accumulation was not, however, to cause a uniform reduction in temperature throughout the world. Equatorial areas remained hot, if perhaps not quite so hot as at present, and they also received a much higher rainfall than at present (Fig. 8). This may help to explain the gradual movement of the people of *homo sapiens* out of the original homeland in eastern Africa to the northern and southern parts of the continent, where the rainfall was less, the tropical forests less impenetrable, and their traditional way of life, which required access to open forest and savannah so that meat- and fur-bearing animals could be hunted, could continue. And, of course, in the rest of the world they then had access to Arabia (not a desert at the time), and then to Asia and Australia, and eventually to Europe.

In fact, it is clear that men penetrated into Europe and Central Asia and China during the Ice Age. *Homo sapiens* reached Europe, for instance, about 40,000 years ago, long after his arrival in Arabia and the Indian regions (Fig. 2), and at least one individual died in Greece even longer ago than that. About the same time China was reached, and Siberia. All these were already cold regions, even glaciated, but at that time they were going through a slightly warmer phase (these are called 'interstadials' by glaciologists – periods between bouts of more intense cold). But the real point is not so much when *homo sapiens* arrived in these lands, but that they stayed in them when the cold returned. This was one of the marks of *homo sapiens*, his adaptability to new climates.

Geographical Change

The greatest geographical changes as a result of the fall in sea level were in Asia. The Persian Gulf vanished, becoming an area of low-lying dry land connecting eastern Arabia with southern Iran (Fig. 33), a factor which obviously helped in the migration of people eastwards. Further east, the seas and straits and peninsulas and islands of Southeast Asia vanished into Sundaland (see Prologue and Fig. 3); China and Taiwan were attached to each other by dry land, the Philippines were a single huge island, the Japanese archipelago was attached to Korea and so appeared as a peninsula of the continent (with the Japan Sea sometimes a lake, and sometimes connected to the ocean) (Fig. 33). All these areas became inhabited during this period, from about 40,000 years ago. Australia and New Guinea were one island, and Tasmania was part of it – men had reached these areas much earlier, of course (see the Prologue). In Europe the English Channel and the North Sea were dry land as far north as the latitude of Shetland when the cold was at its most intense, though most of it was under the ice sheet in the time of the most intense cold; many of the Mediterranean islands were peninsulas of the mainland, the Bosporos and the Dardanelles did not exist, the Black Sea was a shrunken lake. Alaska and Siberia were linked across the Bering Sea, which was dry (Fig. 20). Some of these new lands were extensive enough to be given their own names – 'Sundaland' and 'Sahul' are examples; in the dry North Sea the land has been given the name 'Doggerland', after the Dogger Bank, which was an area of low hills at the time, or sometimes 'Northsealand';

'Beringia' is the name given to the land joining Alaska to Siberia, from the Bering Sea which now covers it. It has been concluded that about 25 million square kilometres of extra land appeared as a result of the fall in the world's ocean level; but perhaps just as much, or even more, had vanished under the ice caps.

The enlargement of the ice caps had major effects on the world's vegetation. In brief, the successive bands of the different climatic and vegetation zones around the world were squeezed, being pushed north and south (figs. 8 and 9). In Europe, for example, most of the land as far south as the Alps and Pyrenees was tundra, with a permanently frozen subsoil ('permafrost'). In this condition the summers were cold, and the winters frozen, so that snow covered the ground for six months or more every year (Fig. 10), and no trees would grow. Spain and Italy and the southern Balkans were lands of coniferous forest, like modern northern Scandinavia. The Mediterranean, smaller and lower than now, was cooler, and south of it were deciduous forests (such as now in Europe north of the Alps) along the North African mountains.

The Sahara Desert was affected by all this, but only intermittently. For much of the last Ice Age (from about 60,000 to about 12,000 years ago) the Sahara was dry, but the desert areas were much smaller. The fluctuations in the general climate produced extended periods of rain and so rivers flowed and there was a great lake in the centre, of which Lake Chad is now the remnant, and there were other lakes in several places from the Nile to the Atlantic. The broad area of desert which now lies south of the North African mountains was therefore much narrowed, and the savannah grasslands comprised about half of the modern desert areas were pastures for large numbers of herbivorous animals, which were therefore also prey for their carnivorous enemies, including man. One of the wet periods began a little before the end of the Ice Age, and was coming to an end by about 8,000 years ago, with a return to dry conditions. The desiccation process will have begun earlier than this and lasted a long time, so that the desert gradually expanded southwards (see also Chapter 6 and figs. 50 and 51 for the effects of these changes).

The increase in the rainfall in the tropical regions had its effect also to ensure this expansion of the savannah regions, and a relatively small expansion of the tropical forests further south took place. The fluctuations of the rainfall regime therefore meant that much of the great hot desert areas might disappear for several thousands of years at a time; the reduction of moisture in the atmosphere when the cold returned then brought the deserts to expand again; the last few millennia of the 'last' Ice Age was such a time, though, as will be discussed later (Chapter 6), this is a much over-simplified account of what really happened.

Further east the monsoon region of the Indian Ocean was pushed southwards, and perhaps weakened, depending as it does on heat, so that India during the Ice Age was an arid land, though the ice in the Himalayas fed strong rivers in the north in the Indus and Ganges river systems, as it does now. (For the same reason the incidence of hurricanes in the mid-Atlantic, and of typhoons in the Pacific was much lower than now.) China similarly was drier than at present, and in Australia the arid centre expanded well beyond its present size; the people there needed to use the lands which were exposed by the lower sea level outwards from

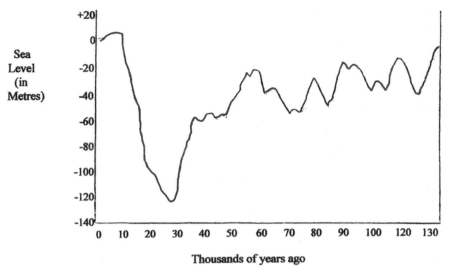

Fig. 7. CHANGES IN SEA LEVEL. Fluctuations in the level of the sea over the recent Ice Ages reflect the quantity of moisture held in solid ice, and so indicates the depth or otherwise of the Ice Ages. Note that the present warm period has lasted ten millennia (top left of the diagram); only one comparable warm period has existed since the evolution of *homo sapiens* – that which allowed human beings to reach Australia. And the process of the sea level rising to its present level took well over 10,000 years to accomplish.

the present coast, but some of them had certainly worked out a way to live in the desert part of the continent (Fig. 11).

The temperate regions of the modern world – Europe, Canada, Siberia – became the new deserts, but of the ice variety. They were partly under the ice, and nearby regions, such as central and western Europe, was so close to the ice that the land was frozen for much of the year. There and in North China the dry cold encouraged the formation of loess, in which the dusty wind-driven soil blown from dry lands were formed into thick layers in the area south of the ice; there were large areas of this in Eastern Europe but even greater areas in China. (This condition must have been very difficult to endure at the time but, in the usual historical irony, this loess land became very fertile once the Ice Age ended – just add water.) This area of cold desert extended eastwards all the way through Siberia to Alaska, a mixture of ice caps, especially on frozen mountain ranges, flooded areas, and frozen tundra.

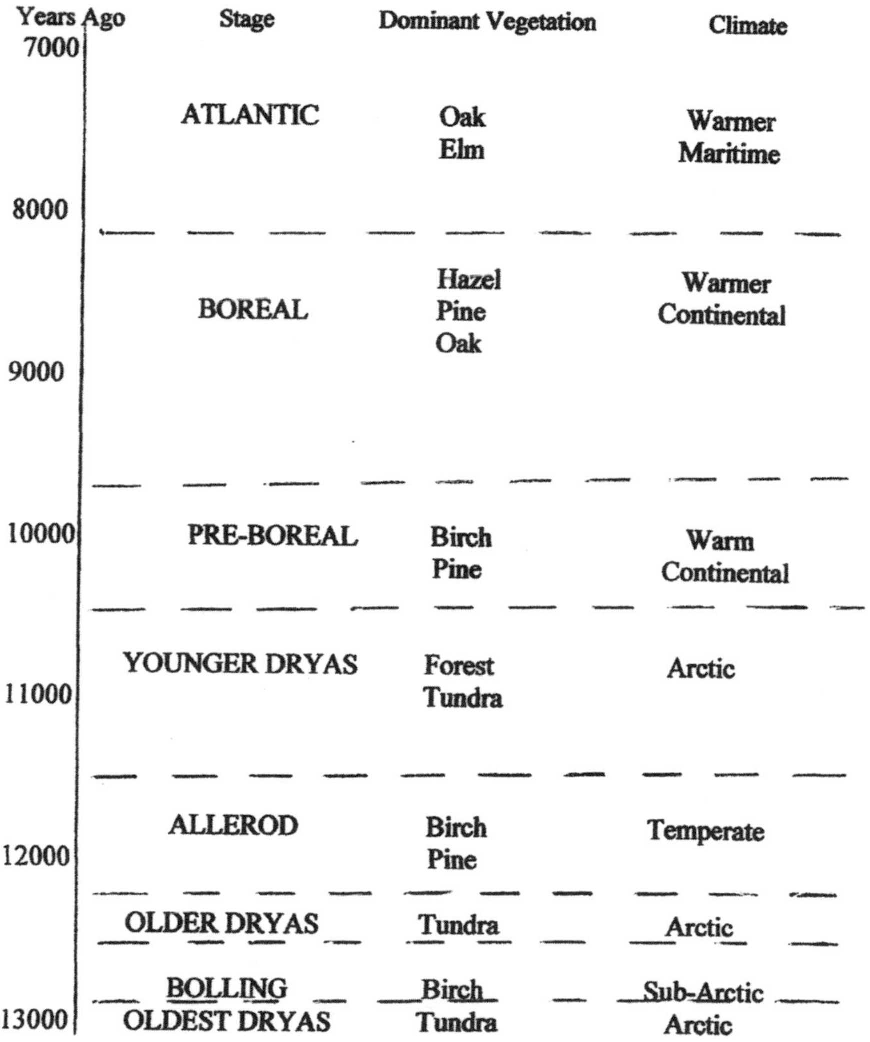

Years Ago	Stage	Dominant Vegetation	Climate
7000			
	ATLANTIC	Oak Elm	Warmer Maritime
8000			
	BOREAL	Hazel Pine Oak	Warmer Continental
9000			
10000	PRE-BOREAL	Birch Pine	Warm Continental
	YOUNGER DRYAS	Forest Tundra	Arctic
11000			
	ALLEROD	Birch Pine	Temperate
12000			
	OLDER DRYAS	Tundra	Arctic
	BOLLING	Birch	Sub-Arctic
13000	OLDEST DRYAS	Tundra	Arctic

Fig. 8. **CHANGES IN CLIMATE AND VEGETATION.** The present warm period began with considerable fluctuations in the climate, and so with similar changes in vegetation. The chart shows the changes which took place in north-western Europe over the period from the beginning of the warming (the Bolling period) to the beginning of the settled climatic period (in archaeology this is the Mesolithic period). It was only with the warming of the end of the Younger Dryas that the vegetation cover of coniferous and deciduous forest replaced the tundra – and only then were men forced permanently into a new life.

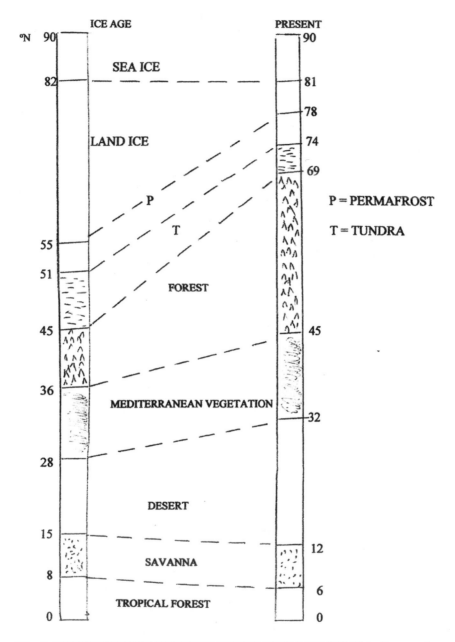

Fig. 9. NORTHERN HEMISPHERE VEGETATION ZONES. This diagram shows the changes in vegetation zones from the North Pole to the Equator in Europe and Africa brought about by the end of the Ice Age. It will be seen the greatest changes are in the replacement of land ice by forest and the expansion of the desert belt (that is, the Sahara). Other vegetation zones were relatively less affected in size, though they shifted northwards.

North America was an even more drastically changed land in its geography than Eurasia (Fig. 13). Large ice sheets completely covered Canada, and south of that there were enormous lakes along the edges of the ice; the present United States was colder and drier than they are now, though the tropical areas were probably little different (Fig. 24). South America was probably the least affected of all the great continents by the Ice Age, gaining a little land around the coasts as the sea level fell, and with a relatively small ice sheet in the far south, and on the Andes' tops; but the rest of the continent was warm and humid. (See Chapter 3 for more on this.)

Migrations during the Ice Age

This, therefore, was a very different world for its inhabitants than the one with which we are now familiar. For people alive when the Ice Age began to relax its grip between 14,000 and 12,000 years ago – necessarily slowly at first – there had never been any other climatic regime, nor was there likely to have been any racial or ancestral memory of anything else. They and their ancestors for 50,000 years had known the world only as it was then, though there had been fairly generous temporary fluctuations; yet few of the fluctuations had made a serious dent in the ice. Measurements by ice scientists have indicated that it is only since about 12,000 years ago that the global temperature has risen to its present level, which we rather complacently think of as 'normal' or 'optimal'. In fact, of course, for much of human historic time, the 'normal' temperature was several degrees lower than it is now, and the Ice Age was 'optimal'.

For the humans in Africa the Ice Age was a benign period, a little cooler than at present, with great extensions of the savannah lands which were their favourite preferred hunting grounds, where there were large herds of huntable animals for their prey. This was also the period when the colonists of Australia spread throughout the continent, moving into every corner of it (and into the other lands of Sahul), into Tasmania, and even into the desert centre, as well as the more welcoming coastal areas and the south.

When the last – or latest – phase of the glaciation (the Last Glacial Maximum) began to fasten its grip on the northern lands, from about

15,000 years ago (the largely oceanic south can be ignored in all this, though the Antarctic ice cap grew and spread mightily) humans had already spread successfully through the hot lands from Africa as far east as Australia, but they had not gone very far towards the north, blocked by the ice caps. A warm period then allowed, or encouraged, movement northwards around 40,000 years ago. It is a paradox, therefore, that it is just in the Ice Age period, as the ice caps grew and the permafrost spread, and the more benign climatic regions moved southwards and were compressed, that the humans whose ancestors had moved north into Europe and Siberia and China during the preceding warm times, decided to stay there, and made a successful life amid the ice and the snow, throughout the cold times.

Just as the Australians were adapting to the deserts of Central Australia, and the Melanesians were developing their seafaring skills to create a seaway network of continuing contact among the islands east of New Guinea, therefore, their distant cousins in the northlands were adapting to the intense cold of Europe and to the equally intense cold of China. Further, they also spread northwards to live for at least part of the year at the very edges of the ice caps (see Chapter 2), and in the eastern Asian lands to the north of China they infiltrated between the discontinuous ice caps of the Far Eastern mountain regions as far as Beringia and Alaska – where they were eventually stopped by the huge North American ice sheets, but where, as will be noted later, they also contrived a successful life. In fact, it may be that some had passed south into the American continents even before the great North American ice sheets blocked the way; this is a controversial matter (see Chapter 4), but there seems no obvious reason why contemporaries of people who were expanding into Europe and Australia and taking to the Melanesian seas should not have moved into America.

This is a remarkable record of adaptation – the same human animals were able to organise their lives successfully to live in deserts, savannahs, islands in the sea, and in tundra close to the ice deserts. They lived in these conditions for 40,000 years, while the ice caps dominated the northern lands, and some of them successfully managed their lives amid ice and snow and the permafrost. Then, from about 14,000 years ago, the Ice Age began to relax its grip ever so slowly – the Bolling 'warm' period.

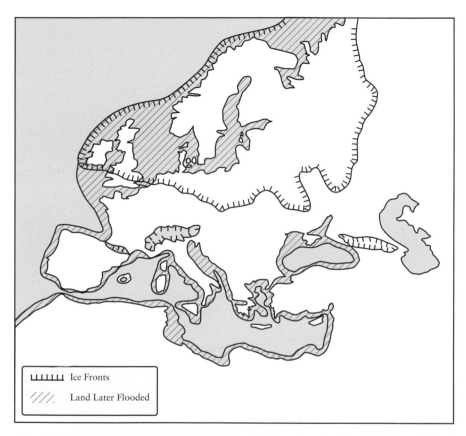

Fig. 10. EUROPE IN THE ICE AGE. At the height (or depth) of the Last Glacial Maximum, when the climate was at its coldest, about 20,000 years ago, ice sheets covered perhaps half the continent. Note especially, apart from the ice, that permafrost covered much of the rest of the land, certainly as far south as the Alps and the Danube, rendering the land unfit for growing anything other than mosses and low growing shrubs – which was suitable food for the cold adapted animals such as mammoths.

Human beings were now to be affected by this new challenge to their proven capacity for adaptability.

There is, of course, more to this human adaptability than simply being able to live with the cold or the heat or the dry. The people who adapted had to do so by working out new ways of living in a different climate, hunting different animals, eating different vegetable foods. To do this they had to find a means of sheltering themselves from the intense cold of the European Ice Age winter, or from the searing dry heat of an Australian desert summer. Caves were their usual refuge in both of these climatic

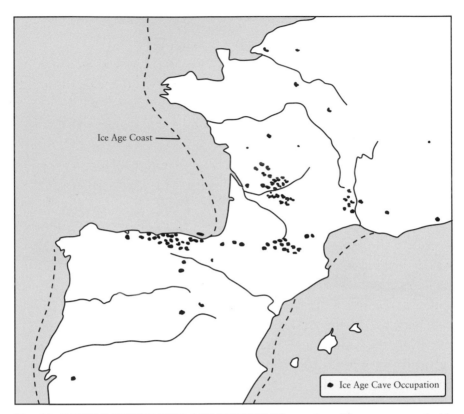

Fig. 11. CAVES IN USE IN ICE AGE EUROPE. The tundra of Europe swarmed with animals which were the prey of the small human population. This was the central element in human life in the Ice Age, and it is reflected in the art practised in the homes inhabited by the hunters. We do not have examples of their artistic lives in their summer tents, but in their caves they left the full repertoire of the animals they hunted. These cave paintings are remarkably accurate anatomically, but one would expect hunters to know what the animals were like. They also show admiration and even affection for the animals at times.

For us, the pictures provide an insight, partial no doubt, into the ideas and attitudes and life of the people living in Europe in the Ice Age and after (Gallery II). The animals they drew and painted and inscribed and sculpted were those which they knew as neighbours and prey. They showed bison, horses, fish, reindeer, mammoths, cattle. They show themselves hunting these animals and incidents when the animals won the fight. But these pictures tended to be made in the deepest and darkest parts of the caves; they were not producing a picture gallery to be admired.

regimes, but in the summer in cold areas, or in winter in hot lands, they came to use tents.

Weapons were needed which were capable of killing great animals several times the size of a man, or could move several times faster than a man could run. In more than one place bows and arrows were invented, traps developed, and in more than one place the first dogs were domesticated to assist in the hunt in return for an assured food supply, exchanging their wild freedom for the certainty of food and warmth and shelter and human company. The Ice Age, in whatever part of the world man had reached, had been a time of constant, if slow, experimentation and invention. The people who had to cope with the new conditions of a warmer globe as the temperature rose were themselves already adapted to inventiveness and survivability: the new conditions were a challenge, not a threat.

The changes in the global climate therefore, came upon a population of hunters and gatherers who, in Europe and northern Asia at least, had become habituated, even adapted, to the cold climate. Their skins had paled, their nostrils had narrowed to slow the entry of cold air; they were clothed in the furs of the animals they had hunted for food, their summer tents were made from the skins of other hunt victims; they used a variety of specific tools as well as several weapons, and their repertoire of both had expanded and specialised over the cold millennia; they had invented, for example, the needle, which enabled them to manufacture the well-fitting clothes necessary to survival in intense cold, and to make spacious tents for protection in the cold (and bags to carry things in). Another invention was the bow and arrow, which enhanced their ability to hunt the animals at a safe distance, and which therefore provided them with food and furs and bone. They lived during the coldest times of the year in caves, whose walls and roofs they painted with pictures of the animals they hunted – for, like the early inhabitants of Australia, they were artistically inclined. They had, that is, developed a useful and generally comfortable style of life, and they were skilled and competent hunters, artists, and manufacturers.

This is not to say it was an altogether pleasant life. The hardship of the cold was unremitting, hunger was a constant threat, and the cold could frequently be intense and killing. Few of the people ever lived for more than forty years, and most of them died before they were thirty. They were superstitious, hence their paintings were done in dark corners where they were probably never seen except by the artist and the spirits of the animals they hunted, and they remembered their dead, burying family members with reverence. The conditions they lived in enforced careful forward planning. In the depth of winter hunting was impossible, so that food collected during the warmer seasons had to be stored and consumed with careful economy – the cold outside the cave will have preserved their meat.

This was not, however, a culture frozen into immobility. I have noted already the invention of such useful tools as the needle and the bow, and both of these emphasise how small inventions like these may be the most useful. The bow, for instance, also required the development of efficient and light arrows, made from carefully selected wood, and the improvement of the arrows' penetration power involved the use of small sharp pieces of stone fastened to the arrow point. These are called microliths ('small stones') by archaeologists, and their use was another development of the Ice Age; in fact, it was these tools which have become the defining characteristic of the subsequent Mesolithic period for the archaeologists. Such stones were also used as spear points, and in some cases several small sharp stones were inserted to form a serrated edge to make a nasty club. Flint and, where it could be obtained, obsidian, were the best stones to use, for both kept a sharp edge for a long time, and so another summertime task was to locate a flint source and gather supplies for later manufacture. These people became practical geologists, and experimentation in the manufacture and fitting of microliths went on for millennia. The eventual result was a weapons-kit of steadily increasing efficiency. Men could now hunt with a much lighter set of equipment than before, and more effectively.

It was the conditions of Ice Age Europe (the 'Palaeolithic') which pushed these developments, since a regular food supply in the enervating cold was even more essential than in the warmer climates, where fruits and vegetable foods were more easily available the year round. The enforced immurement of winter provided time during which inventions and clothing sand new tools and weapons could be made, and paintings could be done. And yet, give or take a few tools and clothes, the same increase in efficiency can be noted among their cousins in Africa and Asia and Australia, so it was not just an effect of the Ice Age on the people in the north. How the people of these various regions, and those elsewhere who were not so strongly affected by the change in temperature than

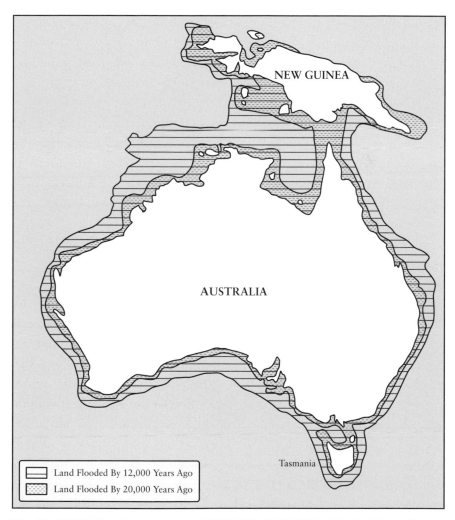

Fig. 12. THE SHRINKING AND DIVISION OF SAHUL. The familiar outlines of the islands of Australia, New Guinea, and Tasmania are blurred by the Ice Age coasts. The outer line marks the coast at about 20,000 years ago, the coldest part of the Last Glacial Maximum. The inner line is the coast at about 12,000 years ago, not long before New Guinea and Tasmania became separate islands.

those in the north, but were affected by other climatic changes, were able to cope with the change is what I am looking at in the rest of this book. The adaptability of the species has been well demonstrated – it existed from its origins, as decorations in South African caves of a hundred millennia ago, and the crossing to Australia, and the successful lives of hunters on

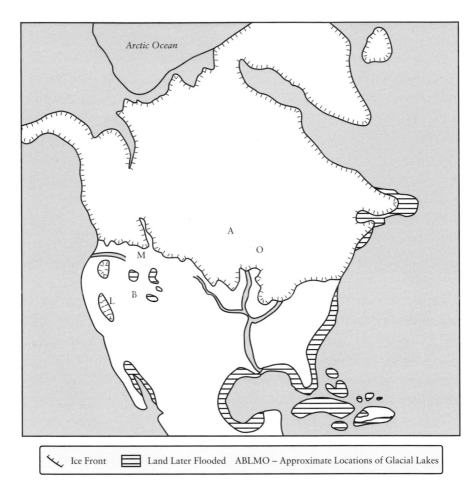

Ice Front   Land Later Flooded   ABLMO – Approximate Locations of Glacial Lakes

Fig. 13. NORTH AMERICA IN THE ICE AGE. The two great ice sheets, the Laurentide and the Cordillera, reached out to link up at the coldest time of the Last Glacial Maximum. The approximate locations of the largest ice-dammed lakes are shown, but they tended to fluctuate in both size and location. (A – Agassiz; B – Bonneville; L – Lahontan; M – Missoula; O – Ojibway). These lakes, of course, tended to appear and enlarge when the ice was melting. The great spillways are also shown – the modern Columbia, Mississippi–Missouri–Ohio, the Hudson, and the Saint Lawrence.

the European tundra all show. The next challenge required this quality of invention and adaptation to be exercised even more ingeniously to do more than invent new weapons; it was to be necessary to invent new ways of living.

Further Reading

A very useful description of the action of the ice is in Ian Cornwall, *Ice Ages, Their Nature and Effects*, London 1970; for much detail on climate one may consult Lamb, *Climatic History* (see Chapter 1). An up-to-date discussion with notices of the latest research and presented in a most readable way is in Burroughs (see Chapter 1), which also has good references to more detailed researches. The Toba eruption is described by S. Romero and S. Self, in *Nature* 359, 1992; the Laki eruption is in 'The Summer of Acid Rain', in *The Economist*, 22 December 2007.

The expansion of the human species before and during the Ice Age is discussed by Clive Gamble, *Timewalkers, the Prehistory of Global Colonisation*, Stroud, Gloucestershire 1993, though various developments have rendered it somewhat out of date in some details already.

The conditions of life in the European Ice Age (the 'Palaeolithic Age') are usefully summarised in *The Oxford Illustrated History of Prehistoric Europe*, edited by Barry Cunliffe, Oxford 1994 – Chapters 1 and 2 by Clive Gamble and Paul Mellars respectively – and by Gamble in *The Prehistoric Societies of Europe*, Cambridge 1999. Desmond Collins, *Palaeolithic Europe, a Theoretical and Systematic Study*, Tiverton, Devon, 1986, is very technical; see also some of the essays in Paul Mellars (ed.), *The Early Post-Glacial Settlement of Northern Europe, London 1974*. The adoption of the bow is discussed by J.-G. Rozoy, 'The Revolution of the Bowmen in Europe' in Clive Bonsall (ed.), *The Mesolithic in Europe, Papers presented at the Second International Symposium, Edinburgh 1985*, Edinburgh 1989, 13–28.

For Australia see the books by Josephine Flood noted in the Prologue, and the articles by Peter Kershaw, 'Environmental Change in Greater Australia', and Douglas Edwards and James F. O'Connell, 'Broad Spectrum Diets in arid Australia', in the *Antiquity* supplement on *Transitions* (see the Prologue); for India consult D.K. Chakrabarti, *Oxford Companion to Indian Archaeology, the Archaeological Foundations of Ancient India*, New Delhi 2006, Chapters 3 and 4; for America see parts one and two in F.C. Pielow, *After the Ice Age, the Return of Life to Glaciated North America*, Chicago 1991, and the references noted in Chapter 3 below.

Gallery II

Europe in the Ice Age

The habit of painting on the walls of caves in Europe in the Ice Age is difficult to account for; it was certain judging by the subjects of the paintings, largely concerned with the animals the artists hunted in their working time, though what the purpose of the paintings was is not known. It is theorised that, since the painting was done in inaccessible parts of the cave, it was a sort of sympathetic magic, by which the animals portrayed would be more easily hunted, but this is no more than a modern theory. What is certain is that the artists knew their animals very well, and portrayed them exactly, almost as individuals. One cannot help feeling that the artists were painting because they were artists, whether or not they believed in sympathetic magic.

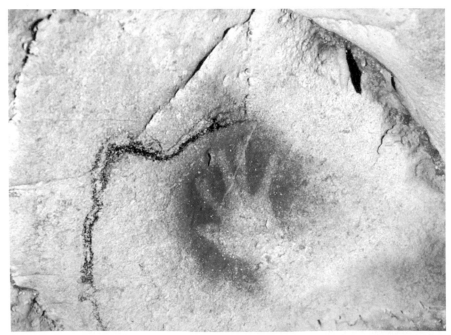

II.1 Chauvet Cave, hands. One of the most common motifs in the painted caves, wherever they are, is silhouettes or paintings of hands – the instruments of life.

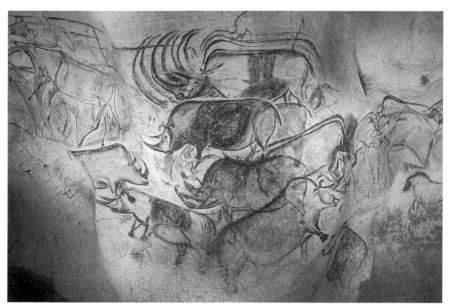

II.2 Chauvet Cave – the animals hunted by the Palaeolithic inhabitants.

Not all pictures are easy to understand. Those of bison show the animal in all its power and that from Altamira is clearly admiring (II.3). But animals to men who made a living by hunting them were also prey. The two other bison pictures show animals that have been hunted.

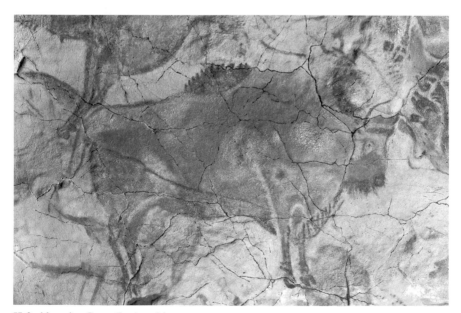

II.3 Altamira Cave, Spain – bison.

Two questions arise over both pictures. In the first is the scene a composition showing an event? (II.4) Is the second a celebration of an actual successful hunt, or as some would have it, a painting of an event which he hoped would come to pass, perhaps by his own prowess as a hunter? Both theories are held by archaeologists, but the second has an air of disbelief that mere Stone Age hunters could actually succeed in killing a bison of such strength and power.

II.4 Lascaux, France – bison. A man falling, perhaps gored by the bison, and a 'standard'.

II.5 Niaux Cave – bison hunted to death.

Perhaps rather more accessible prey were horses and cattle, and both are commonly drawn and painted on the cave walls and roofs. The painting from Altamira (II.6) illustrates the complexity of the cave pictures.

II.6 Lascaux Cave – horses; the original painting showed a cow; the horses were painted over it.

This same complexity is visible in the Lascaux painting (II.7) but the images of horses is drawn as much to show the expertise of the artist as his hunter's skills.

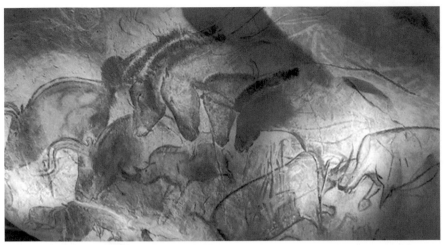

II.7 Lascaux – horses, rhinoceros and cows – a sequence of over-painting.

II.8 Lascaux – three reindeer depicted swimming. In some cases, such as this, one can see no more in the picture than the artist recording a past event he felt worth showing. Men with such artistic skills were surely proud of them and constantly used them.

Chapter 2

Following the Ice

As the climate in the north changed from cold to cool and then to warmer, albeit slowly and discontinuously, the obvious way for migrating hunters and gatherers to cope with the change was to continue living in the same way and do the same things but move to a different place. This chapter is therefore devoted to some communities which attempted to maintain their former lives without changing them, except by moving to a new area – and even that change might be only minor, for nomad hunters were accustomed to move regularly to new campsites and new regions. Since it was obviously the groups who lived in the cold climate regimes who were the most directly affected, it is those who will be singled out here. Others, affected in different ways and reacting in different ways, will be the subjects of later chapters. This is the import of the title of this chapter, in that some communities simply followed the retreating ice and went on living as before.

The New Conditions: Europe

In Europe the change of climate and landscape was particularly drastic (figs. 10 and 16). The ice at times came as far south as Midland England and Ireland, and well into northern Germany; all Scandinavia was covered by an ice sheet to an enormous depth, and this stretched deep into European Russia. South of the ice sheet, from France to the Ural Mountains, the land was tundra, with much of the soil permanently frozen below the top few inches of an intermittently-thawed surface layer. The Alps and the Pyrenees were covered by caps of ice and these smaller ice caps blocked the way from the tundra of central Europe and France towards the Mediterranean to the south. Spain and Italy and the Mediterranean lands were cold, partly steppe land, and partly forest, rather as northern Scotland and Scandinavia are now.

The lower sea level, combined with the ice cover, meant that the coastal geography of Western Europe was wholly different than at present. The southern half of the North Sea was land ('Doggerland' or 'Northsealand') (Fig. 16), with a substantial lake occupying the southern part for a time, formed by meltwater from the edge of the ice to the north, dammed up behind the ice and reinforced by rivers flowing into it out of the south. The English Channel was dry land, with great rivers flowing along it, fed partly from the land, and partly by outflow from that southern North Sea meltwater lake which broke through the chalk bridge which linked the Weald and Artois and eroded it to form the cliffs of Dover and Calais. (This lake did not survive for long once the ice had retreated a little way; suddenly there will have been a way for the waters to drain into the ocean, and it will have flooded out quickly.) The coast of France south of Brittany was pushed well out to the west, leaving a much smaller Bay of Biscay; the new Atlantic coastline lay across the Channel mouth, so Brittany, Cornwall, and Ireland were all a single land and were all part of the continent.

All the territory of Eurasia, from the Atlantic coast to the Pacific, was technically dry land, but it was also frozen land, except briefly in the summer, and was totally frozen and snow-covered in the winter. To the east of Europe the geography was also different. The Caspian and Black Seas, originally both inland seas at the depth of the cold time, were now enlarged by meltwater from the Scandinavian/Russian ice sheet and from the mountains (the Caucasus Mountains were ice-covered) and were themselves tenuously joined together as one sea. These inland seas, and the ice caps on the mountains, virtually isolated Europe from the rest of the world, with only narrow connections by land linking the two.

As the ice melted, therefore, the bands of the climate zones shifted slowly northwards (Fig. 9). The ice caps on the mountains – the Pyrenees, the Alps, the Caucasus, perhaps the Carpathians – slowly retreated, and great volumes of water poured powerfully down the hillsides, scouring out the valleys and spreading the debris over the lower lands. The lowland forests gradually crept up the mountain slopes, just as they gradually colonised northwards as the ice retreated and the tundra warmed. The climate gradually became warmer over the long term, but this was not a dependable and smoothly-continuous development. It was not the case that the ice melted a little bit every year: instead the ice retreated in an intermittent way, sometimes melting with some speed, and at others not at all, perhaps for decades, and even re-advancing considerably for a time. Just as now, years could go by with no change, or with the ice retreating or advancing, and it was all quite unpredictable.

The major changes took centuries to be obvious; they have usually been given specific scientific names. The final period of glaciation is termed the Last Glacial Maximum (a rather optimistic name, perhaps), but other names are given to colder or warmer episodes within the whole, notably the initial warming period which is the Allerod interstadial, and a major revival of cold, the Younger Dryas. When the retreat of the ice restarted at the end of the Younger Dryas, about 10,000 years ago, it would be years before any really appreciable difference in the boundary of the ice became noticeable.

Added to the changes brought about by the rise in the sea level as the ice melted, and the appearance of new land from beneath the ice, was the phenomenon of 'isostatic recovery'. The sheer weight of frozen material in the great ice sheets depressed the land on which they sat; as it melted, so the land slowly rose. The combination of intermittently rising land and intermittently rising waters, neither of which was coordinated with the other in any way except the most general, produced constant fluctuations in local sea levels. In the West of Scotland, for example, the effect of the melting of the ice was to lower the sea level, because the land lifted more quickly than the sea level rose; then the rising sea level took over and raised the local level almost ten metres higher than the present level; finally, the sea level stopped rising, and this allowed the relative change to reverse, so the land 'rose' until sea and land achieved the approximate equilibrium of the present relationship (the change, however, is not yet ended, though it has slowed). (See, as an example, Oronsay, Fig. 44). Note that the term 'recovery' implies that the present situation is 'normal', whereas the cold of the Ice Age is perhaps the 'normal' condition.

The crucial time for archaeologists studying this period is the time of the lowest sea level (between 18,000 and 9,000 years ago), but the later rise has flooded many of the areas where the earliest inhabitants might have lived and has hidden any evidence of their lives. These changes are different in sequence and size in each local region, so that each area has to be investigated for the local sequence of changes. (The incidence of flooding is virtually universal, and clearly lies behind the biblical story attached to Noah: there seems no need to find a specific flood to account for it. Everyone living anywhere near the coast at any time in the period following the global warming at the end of the Ice Age will have known and/or seen land being flooded; it was clearly graven into the collective minds of their successors.)

The New Conditions: Siberia

Europe has been more intensively studied than elsewhere, except perhaps for Japan and maybe parts of North America, and the conclusions reached are not necessarily wholly applicable to other regions, though the major advances and retreats of the ice, and so the major climatic periods, are well known and can be identified all over Eurasia, in each case with its local name. To the east of the great Scandinavian ice sheet, in Russia and Siberia, conditions were not quite as extreme as under or near the ice. Ice caps covered the several Siberian mountain ranges, of course, as they did the Alps and the Pyrenees and the Tibetan plateau, but the huge areas of land between those frozen wildernesses were tundra and were free of ice in the summer. In that sense these areas, which are subject to savage winters today, have not changed as much as Western Europe has. The Arctic ice encroached into northern Siberia (though some parts of the Arctic Ocean may have been open water at times) and large areas of the eastern part of Siberia and the Russian Far East – the Verkhoyansk, Chaskiy and Kolyma ranges, and Kamchatka, for example – were covered in ice, as also were the great Central Asian mountain ranges. But the ice caps were not continuous, and so they channelled the movements of the migratory animals into particular migration routes and to particular grazing areas through fairly narrow passages, and that is where men hunted them. This geography had its effect later, when Western Europe warmed decisively; as a result, Siberia became a refuge for animals which disliked the warmer climate, and for men who continued to rely on those animals. But the Russian Far East was deeply inhospitable to man, and to the animals he hunted in the depth of the Ice Age; archaeological investigation tends to suggest the whole area was uninhabited until conditions warmed somewhat.

Life Changing – into the Mesolithic

For the people living in Europe in the Ice Age, life was apparently quite bearable, despite the cold, for they had largely adapted to it. There were not many people in the continent, for a start, so what resources the land provided were plentiful in relative terms. In the lower mountain lands, particularly in limestone areas, there were many caves whose internal temperature, while not particularly high (usually a uniform 13°C), was at least higher than the winter temperature in the open air; and in those caves the families of the hunters were well sheltered from the winter weather; many of these caves have evidence of long human habitation, though only the best decorated, such as Lascaux, Altamira, and Chauvet, are at all well-known. There are perhaps sixty or more such caves in northern Spain and southern France. They were, that is, utilised throughout the Ice Age, quite possibly by successive generations of the same families (Fig. 11; Gallery II).

The resources available to these hunters were considerable, above all in the form of mammoth and reindeer, but they also hunted bison, bears, wild oxen, aurochs, deer of several sorts, and the smaller animals. These animals are migratory, and in the context of Ice Age Europe this meant that they moved northwards in the spring and southwards in the autumn, changing their grazing lands with the seasons. These were herd animals, in many cases, and they migrated in herds. The frozen tundra winter was unproductive of grazing, though many of the animals dug through the snow to get at their food, a practice easier in the southern tundra than near the ice; the summer, which was fairly brief, rather like that in modern Siberia, was a time when the snow vanished, plants grew and matured swiftly, and animals grazed in order to accumulate fat for the leaner winter. And all the time the men hunted them, moving as the herds moved.

In the autumn and early winter, as the land froze again, many of the animals moved south, to grazing lands closer to the Mediterranean, and the hunters came south with them. It was in this cold time of the year that the men inhabited the caves of southern France and northern Spain, whose illustrated walls have so astonished us with their paintings and carvings and engravings of animals – the very animals which they hunted, and which they knew so well.

Since the animals they hunted were often migratory, it was necessary, in the summer in particular, for the hunters to move out of their caves, follow the herds as they moved north, and to live out on the open tundra. Archaeological evidence has been discovered demonstrating this. In several parts of central Europe sites have been located where these families camped in the open, living in tents, or even without apparent cover, and therefore, given the climate, necessarily in summer. In some places huge quantities of the bones of the hunted animals littered the ground. The campsite at Dolni Vestonice in the Czech Republic is an example. It is located in the loess region of Moravia, where windblown dust from the drying moraines was being deposited. The camp became buried under several feet of this loess (and it lay on earlier deposits of the same stuff); the excavation produced hundreds of animal bones mixed in with the remains of the camp. The covering of loess shows that it was used during the Ice Age, before the loose ground cover had consolidated into modern soil. Nearly every animal which lived in Europe at the time was prey for these hunters, but of course the larger animals were the more valuable, producing more meat and larger areas of leather. The huge bones of mammoth lie interspersed with those of reindeer and other deer; but the smaller animals, fur-bearing, and often hibernating, were not neglected. This was a camp used, probably by the same family group, repeatedly over a period of several years.

The fur-bearing aspect of their prey was as vital to the hunters as the meat which the animals produced. Even in the summer the whole continent was cold, and so the best way to keep warm was to cover oneself in the same fur which grew on the animals naturally, and to sleep covered by that same fur, if a cave was unavailable. In the camps the signs of tents have been found, in which the long bones of the larger beasts such as mammoth were used to support roofs which were made out of the skins of the hunted animals stitched together. One tent, excavated at Pushkari near Novgorod in Russia, was twelve metres long and perhaps more than three metres high, judging by the dimensions of the hollow which had been scooped out to form the interior, and the marks of the tent poles and pegs (Fig. 14). Such a building would be large enough to accommodate a whole extended family of perhaps twenty people, men, women, children, grandparents, and babies – the interior fug would be rank and tangible but

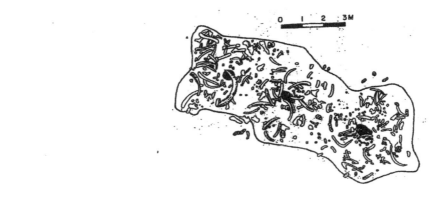

Fig. 14. MAMMOTH HUNTERS' TENT. This is a plan of the excavated remains, and a reconstruction drawing, of a tent found at Pushkari, near Novgorod in Russia. The materials are furs, bones, and tusks, stones and branches. It was over twelve metres long and four metres wide at the base, and it included three hearths (the black areas on the plan), and could accommodate a whole extended family. It may be assumed to be typical of summer accommodation throughout Ice Age Europe, though where it was found suggests that the tent was put up sometime after the ice began to retreat seriously.

warm. The tent also covered three hearths, and it would be some child's job to tend the fire, feeding it with bone where wood was unavailable. Around the tent were scattered such items as mammoth tusks, some of which seem to have been used as tent pegs, holding down the edges of the tents. There were also abandoned and broken tools and weapons, sites of other fires, and occasional burials. Yet all this does not necessarily mean that the occupation lasted very long – possibly no more than a single summer season, though the site might have been used more than once. The debris represents the last day of using the camp; it was material which they did not any longer want. Their winter homes may have been in the Crimea or the Carpathians.

This way of life, sheltering in the caves in the winter, roaming the tundra in the summer, was one which was both satisfying and productive, if tough and hard, and it produced in the hunters a wide knowledge of the European landscape, which they could use to locate their prey and the sites of their movable summer camps. In northern Germany one of their campsites was discovered not far from Hamburg, at Meiendorf, where the bones of reindeer have been found scattered in and around what had been a small lake, beside which the camp was placed. (The site was only reached by digging through a thick layer of waterlogged peat and then examined only by pumping water out continuously.) The reindeer had mostly been killed in the autumn – reindeer antlers grow at a particular rate, and reindeer are all born within a fortnight of each other, so the state of the antlers indicated that this was a summer and autumn hunting site.

Another site close by (at Stellmoor), beside a small lake, had been used for two or three years. Three reindeer skeletons were found well out in the lake, having been deposited there dead but unbutchered, presumably as first fruit offerings when the hunters had arrived. The campsite was established at a slightly different place each new summer, thereby avoiding the messy remains left the year before; and Meiendorf itself was used for about twenty years, or on twenty occasions, probably in succession to the Stellmoor camps.

This was a place to which a particular hunting band had chosen to return year after year for a quarter of a century, knowing that it was a productive area, near to where the reindeer stayed for some time while their calves were born and grew. When the winter approached the hunters would no doubt follow the migrating reindeer herds southwards, carrying such skins and antlers and weapons as they had found most valuable, and in the winter they camped once more in their caves in the southern hills during the coldest weather. These people belonged to what the archaeologists called the 'Ahrensberg Culture', named for a site a little to the east of Stellmoor/Meiendorf. Communities of people of this culture were spread through Germany and into Denmark, and in France and

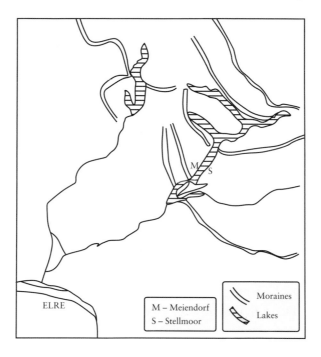

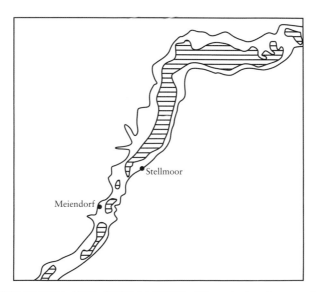

Fig. 15. MEIENDORF AND STELLMOOR. The reindeer hunters lived on the edge of the ice sheet where a series of successive moraines left by the slowly receding ice blocked the stream to form a series of small lakes along the '*tunneltal*'. Both sites were in similar situations but at Meiendorf the killing of the animals was done a little distance from the living site. The main hunting was done in summer and autumn after which the winter would clamp down.

Spain it was called the Magdalenian; the general similarity of the tools and weapons and lifestyle suggests a constant communication across the whole area of these cultures. It was probably characterized by the relationship of a hunting band with a particular reindeer herd – a beginning, in a distant way, of animal domestication. Stellmoor has produced the earliest European evidence for the use of bows and arrows, a more efficient means of hunting than using only spears and stones. Across the dry North Sea – Doggerland – a similar campsite has been found at Star Carr in Yorkshire, where there is evidence of an early domestication of a dog, an assistant in the hunt, though this site is rather later.

These hunters used the product of their prey for virtually all they needed. The furs and skins provided their clothes and summer tents, the bones and antlers became tools – needles, for example – and weapons and fuel, the meat was eaten, the sinews became thread for sewing the clothes and the tent. Other resources were exploited, of course, such as stone and wood, which were utilised in their various ways for tools and weapons, and fire, but they had already located their preferred deposits of stone, of which flint was particularly prized.

This was a life which was largely stable, at least so long as the hunters' prey was available, but it was not unchanging, even if changes came slowly. The main indicator of changes is the development of flint tools and weapons, which became steadily less crude and more neat and specialized over the preceding 30,000 years of the presence of *homo sapiens* in the European Ice Age. The people were, of course, successful hunters, as their continued existence and increasing numbers during the cold period show; they were artistically inclined and fully capable of rendering, in painting and sculpture, anatomically exact pictures of their prey, as hunters and the manufacturers of tools and hunting weapons would be. Their religious beliefs are only conjectural, but they were respectful of the dead, often burying them with clothes and jewellery and valuable gifts – though this is not necessarily a mark of religious beliefs; and like the Lake Mungo man, sometimes they scattered red ochre on the body; it does seem that the people at Stellmoor certainly believed in some sort of god or spirit, to whom the first fruits were offered for a successful hunt.

In all this these Palaeolithic people were essentially like all other Stone Age – and later – peoples until the present day. The skills and beliefs and capabilities which they possessed were used to face the greater changes which took place in their world as the ice melted and retreated, and as the seas rose (Fig. 16).

One of the early geographical changes as the ice melted was the separation of the main Scandinavian ice sheet from that which was centred on the Scottish Highlands. The pent-up southern North Sea lake could then drain away, further eroding the gap in the Weald–Artois chalk ridge. The area which is now the North Sea became an open plain, seamed by rivers and streams, and with frequent lakes. Its geographical heart was a group of low hills, which much later, when submerged, were to be called Dogger Bank, hence the name for the emergent land of 'Doggerland'. Here roamed the animals the hunters sought, and the hunters of course followed them (Fig. 13).

This land had many rivers, large, slow, wide rivers. The Thames ran out into the tundra of the North Sea and joined the Rhine, which flowed southwards to flow along what is now the Strait of Dover and the English Channel through the gap worked by the drainage from the meltwater lake. Other continental rivers flowed north into what is now the northern part of the North Sea – an elongated Elbe and Ems, and a Rhine branch through the hollow of the Zuider Zee. Once the country emerged from the ice, it had only a limited time for existence, as the sea level rose to flood it, but it did last several thousand years, with a shore somewhere between north-east England and a wider and longer Elbe estuary, though this did tend to move with the melting of the ice and the rising sea level. The relative flatness of the region will have ensured that this shoreline was very vulnerable to flooding, and eventually to the slowly rising sea.

Every summer this was a major hunting ground, or so it has been assumed. Mammoth bones, reindeer bones, hunters' weapons, and so on, have all been found in several places, often dragged up by modern fishermen's nets (which was one of the original clues to the land's existence), and recent technological developments have enabled part of this old land to be mapped, hills, lakes, rivers, and all.

The hunters who roamed and camped in these lowlands were of the Ahrensberg Culture, or the Magdalenian, like those who come to Meiendorf; there were no doubt summer camps of the Stellmoor-Meiendorf type in several areas of this land; the 'Dogger Hills' have been

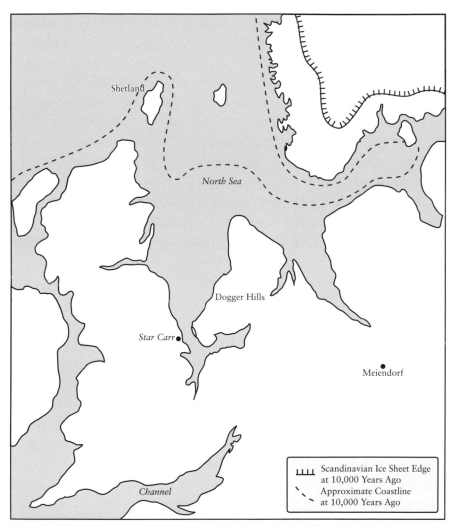

Fig. 16. DOGGERLAND. The geography of north-west Europe as the ice began to retreat about 10,000 years ago, and the sea level rose. A wide expanse of low-lying land became flooded soon after that date, and the people began moving away. By 9,000 years ago some had gone north to Norway, where they would be able to live the same life as in Doggerland. Others will have gone east and west. Star Carr was roughly contemporary with the flooding; Meiendorf is earlier.

suggested as an especially favoured area, and one which, being drier than much of the territory would be the sites of camps and the likely place where regular celebration meetings of the various hunting communities would take place.

The flat nature of Doggerland also implies that the end – the flooding of the lowland and the creation of the North Sea – came fairly quickly once the rise in the sea level had reached a critical point. The 'Dogger Hills' became an island, one which may have remained in existence for a couple of thousand years. The dating of these events is vague, but Doggerland certainly existed as an unglaciated plain by about 13,000 years ago, and much of the flooding had taken place by about 10,000 years ago, with 'Dogger Island' surviving until 8,000 years ago (all dating is approximate and liable to be changed, of course). The land therefore existed through much of what archaeologists call the Mesolithic, but had vanished by the time the Neolithic began, about 6,000 years ago.

One community which will have also experienced these changes, and may well have witnessed the encroachment of the sea into the land, was that which was settled at Star Carr, in present East Yorkshire. This was a typical, even archetypical, Mesolithic site. It was a scatter of houses, in effect a village, on the edge of a lake. The people made use of every kind of local material, and were inventive and prosperous. They used boats, they even built a wooden jetty to make waterborne work easier, they made extensive use of wood, not just for their boats, but for their houses and tools and weapons, and they laid down wooden foundations to lift their houses away from the damp ground. They used locally-sourced yellow flint, and they celebrated by donning masks made of reindeer skulls and antlers, perhaps in a dance to attract sympathetic attention from the reindeer spirits for their next hunt. The village is dated to a little after 11,000 years ago, at a time when Doggerland still existed, but when it was already beginning to be flooded. No doubt some of the hunters from Star Carr travelled into Doggerland in their hunts, and perhaps joined in the communal celebrations when the several communities met to exchange news, select wives, and trade artefacts.

Change

The people who lived at Star Carr, at Ahrensberg, and on Doggerland will have become gradually aware that the climate was changing. Events such as the removal of the North Sea lake, the increase of the area of available land to the north as the ice retreated, and the spread of trees northwards into the tundra, showed all this, and the good general knowledge of the geography of the continent which they probably had, will have given them a basic standard of comparison. They will have known what was happening, though what cause they ascribed to it remains unguessable; since such changes had been going on for a long time they may simply have assumed it was normal. Perhaps over the lifetime of a hunter, or maybe the hunter and his son – 30 to 50 years – as the ice melted the new situation will have become slowly clearer. Gradually they will have realised that the new conditions were creating a problem for them. For, as the ice retreated, and the permafrost melted, and the trees advanced northwards, their old life was going to have to change, in one way or another. The people living in Western Europe were more deeply affected than most, simply because the changes were more drastic than in most other areas. The animals they preyed on were moving away, parts of their lands were flooding, new lands were emerging from the ice; in fact, the whole of their world was changing, and their lives were therefore going to have to change as well.

The obvious solution to the problem, and the one which most groups probably adopted at first, was simply to make use of the land which had become available by the melting of the ice, just as the animals no doubt did. So the band which had camped by the small lakes at Meiendorf and Stellmoor, near the future Hamburg, could move further north, into the newly ice-free Denmark or into the still dry North Sea lands, and continue its own way of life of hunting reindeer – which had also no doubt taken advantage of the new land available – with little change. For some time the change was perhaps welcomed, since in the region of their hunting grounds, there was a larger area of land over which to hunt, and in their winter homes in the south the colonisation of the area by trees provided them with useful further resources for fires and wooden tools and weapons. And, of course, it became warmer, little by little, which was no doubt something to be welcomed.

But the hunters could not continue simply moving northwards. One of the results of the withdrawal of the ice was the flooding of the depressed area freed of the Scandinavian ice; this became the early version of the Baltic Sea. The existence of dry land in Doggerland proved to be only temporary: these low lands became flooded as the quantity of water in the sea increased and its level rose. So the descendants of the band which had used the campground near the small lakes near Meiendorf will have found that its further northward and westward progress was barred by flooded land as well as by the ice which was still in Scandinavia, even if the edge of the great ice cap was now further north.

The bands of people which used to hunt over the lands of the North Sea found that the northern shore was moving southwards, and the river valleys were being widened (and new estuaries also became wider and were deeper and became salty). Doggerland was fairly flat and low-lying, so will have become flooded over only a fairly short time. The net result of all the changes was to restrict the area to which the animals and the hunters could move. In addition, the colonisation of the former tundra by the trees made it impossible for the larger animals, which were precisely adapted to the foods and conditions of the tundra, to continue living there. They moved away to where the new areas of tundra still existed – that is, to the east, and into Siberia – or they died off; the last mammoths in Britain died about 14,000 years ago, a bull and several calves – according, at least, to present knowledge.

This change did not happen all at once, of course, though in the North Sea the flooding may well have been relatively rapid, say within a generation or two. Doggerland shifted from being a wet lowland to being a sea with islands; the Dogger Bank, somewhat higher than every other area between England and the continent, certainly stood out as an island for quite some time after the rest of the North Sea and the English Channel had been flooded. Britain then became an island separated from the continent from about 8,500 years ago.

One result for the hunters throughout Western Europe was that the quantity of meat available from hunting any one type of animal became less, as the larger animals decreased in number, and as the largest of the animals, the mammoths, disappeared. (For the hunters this was, in economic terms, a recession.) Hunters never relied wholly on animal meat, of course, for they gathered roots and vegetables and berries and fruits as well when they could, and, as at Star Carr, they fished the seas and rivers and lakes. The human hunters were not so numerous as to overwhelm the resources. Meat could still be obtained if enough animals could be killed, but it will have become more difficult and more time-consuming to do this.

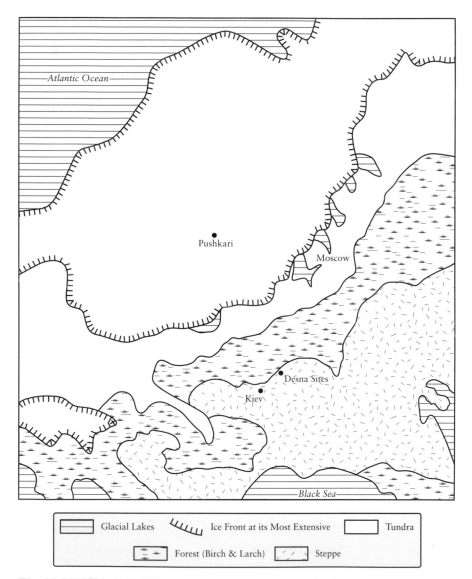

Fig. 17. RUSSIA AND UKRAINE UNDER THE ICE. The ice sheet spreading from Scandinavia dominated the climate and vegetation of all Eastern Europe as well as the west. The hunters lived mainly on the steppe and hunted there and in the thin birch and larch forests. The Desna valley was well placed for summer camps. (Moscow and Kiev are shown for orientation purposes.) Pushkari was the site where the tent reconstructed in Fig. 14 was found.

In the circumstances of the flooding North Sea, with the sea encroaching not only on the northern shore, but along the rivers, the people of the area probably came to rely increasingly on fish for their food. This converted the people into fisherfolk, hunting fish rather than land animals, and this becomes the clue to what happened next in that region. We know from the Star Carr evidence that the people there ate fish, and that islands in the temporary lake, 'Lake Flixton', around which several Mesolithic settlements existed, were occupied – hence they had boats. The flooding of Doggerland, with its lakes and rivers, was the ideal environment in which boats would be used. And it is known that the Hebridean islands were occupied in the Mesolithic period – hence boats again.

Meanwhile in what became continental Europe, especially towards the east, the retreat of the ice had an equally large-scale effect, even if it was not quite as drastic there as the disappearance under water of large areas of land was in the west. The retreat of the ice in what is now European Russia was more rapid than anywhere else so that by 12,000 years ago the ice sheet was confined to Scandinavia (and a small ice sheet continued in Scotland for a time). Most of European Russia was therefore clear, and this includes the separate ice sheets in Siberia (Fig. 17), but it remained a cold region with extensive permafrost and tundra. The advance northwards of the forested belt was thus slower and had less effect in this area. But this was exactly the environment which had disappeared from Western Europe, so that the larger animals, particularly the mammoths and bison and reindeer, which removed themselves from Western Europe or died out there in the greater warmth, or were hunted out, survived for several millennia in the lands to the east.

The New Warming

After several thousands of years of erratically increasing warmth (the 'Allerod interstadial'), between about 12,900 and 11,600 years ago, the climate reverted to a cold phase (the 'Younger Dryas' – more will be said of this in the next chapter), but after that there was another period of fairly steady warming: one of the early benefactors was the Star Carr settlements and neighbouring places. This produced the warmer climate which has now lasted for over 10,000 years (not a particularly long time as Ice Age history is measured), though with the expected fluctuations. The animals which had adapted to the cold conditions, which had lasted for dozens of millennia, moved into the new areas where the food supply was most congenial to them, the mammoths and others to Siberia, whence they spread as far as Alaska, and continued in Siberia for another 5,000 years or so, at least (figs. 20, 21).

The mammoths were the most obvious of these cold-adapted animals, for their bulk produced great resources of fat to cope with the winter cold, and they were clearly unsuited to a warmer climate. Other animals which were equally adapted to the cold included wild horse, reindeer, bears, and smaller beasts such as foxes, Arctic hares, and other fur-bearing animals. These had all formed the prey of the hunters in the depths of the cold times – and in some cases, of each other. As the climate warmed and the animals moved, driven out as much by the spread of the forest northwards as by anything men could do, the hunters were faced with a choice: either stay in their habitual hunting grounds and do without such beasts of prey as the mammoths, or follow the animals.

The retreat of the ice in the north revealed the new geography in both the east and the west. The Scandinavian ice sheet at its greatest had extended south well into northern Germany, where a long series of moraines formed of mixed earth and rocks was left in Denmark, Mecklenburg, Brandenburg, and Prussia to mark its original southernmost reach. The meltwater from the retreat of the ice from European Russia formed the valleys of the great rivers of Russia – the Dnieper, Don, Donets, Volga, and others – gouging their beds deeper by the force of their flow and the weight of water. Further to the east the lands from Poland to Alaska became – or continued to be – the new roaming grounds of the great mammoths.

Hunters of the Ukraine

Animals such as these required large quantities of forage, which could not be located in the forest – quite apart from the difficulty the big animals had in moving through dense undergrowth and close-growing trees, with their visibility much restricted. Some of the mammoth hunters, tied to the migrations of the animals every spring and summer, naturally followed where they went, though they found it impossible to go on living in caves in the more open lands of Eastern Europe. Archaeological remains indicate that they now lived the year round in the same sort of tented camps which had formerly been used in the summer, and that they were still migrating with the animals.

Where the cold–climate animals went, therefore, at least some of the humans who hunted them followed, and continued to hunt them. Sequences of camp sites dating from this period have been located and excavated in the river valleys of Russia and Siberia. The hunting of the area, a land of grassland and tundra, with trees in the better watered river valleys, must have been good. There were strings of sites along the Don River – twenty of them spread over a relatively short thirty kilometres – and another group on the Desna (a tributary of the Dnieper), where there were nineteen sites more or less grouped in a small area (Fig. 17).

These camps were not all inhabited at the same time, nor were they inhabited continuously. We must see them as possible annual camps occupied during the summer months, like those at Meiendorf and Stellmoor, where the principal prey here was presumably mammoths and other large animals of the tundra. Both of the groups of camps mentioned, on the Don and the Desna, were fairly close to the edge of the ice when they were first occupied. Each group of sites might well be the successive camps for the same extended family over a fairly long period of time, perhaps even thousands of years since so long as the animals continued in the area, the hunters could continue to use the same camp site even after the ice had retreated, even possibly eventually using them the year round and not just for the summers. This theory is encouraged by the absence of sites on the open plain between these groups of settlements. It is always dangerous to argue in this way, since it is essentially both an argument from silence and contains a lot of assumptions, but the sites discovered do all tend to line the banks of rivers, which would make sensible places beside which to camp.

The migratory animals moved away as winter froze the land, and the hunting people will generally have moved off with them. The camping grounds by the rivers of the Ukraine were probably only their temporary summer arrangements, like Meiendorf and Stellmoor. The climate continued to get warmer, the ice went on retreating, and the great animals perforce migrated steadily further north and east, and were followed by the hunters, until they disappeared from the Ukraine and western Russia altogether, possibly having been hunted out. Given the size of their food requirements, the population of mammoths cannot ever have been very large, and the removal towards the east would only be gradual. Whatever the precise causes of their extinction in Europe, the ultimate reason was the global warming of the end of the Ice Age. The same happened to a suite of other cold-adapted animals such as the woolly rhinoceros: others, such as the reindeer, Arctic Fox, and so on, which required lesser quantities of food, survived and continued in the northern lands.

Solutions – Following the Animals

The movement of the hunters north and east in pursuit of the great beasts was a viable and sensible response to the effect of the changing climate. The hunters' way of life was predicated on killing the great animals who lived on the tundra, so that where the animals went, the hunters had to go also, since they were, just like the animals they hunted, migratory. Their movements to north and east were no hardship, and marked no real change to their lives. Their camp sites were equally moveable, and new ones could easily be established. But the mammoth hunters did eventually find that their prey, their livelihood, disappeared. It took a long time, but they were destined for extinction also, at least as hunters of mammoths. No doubt, being human beings, they adapted in the end to this new situation by learning to hunt other animals, for they had never wholly relied on one type of animal, but a single mammoth, preserved in the ice, would probably feed a family for weeks. Their ultimate descendants are the tribes of hunters whose remnants inhabit northern and eastern Siberia at the present day.

In Western Europe some fairly large animals survived when the mammoths left, bears, deer, and reindeer, and it was quite possible to continue relying on hunting for a living. But as the forests spread it obviously became more difficult. The reindeer in particular kept moving north with the ice, and the remaining available prey were generally smaller in size and hence yielded less meat. Hunting in the forest is clearly a much more difficult operation than doing so in the open steppeland or tundra, though the use of the bow made a close stalk unnecessary. So while the weapons were more useful, the work was harder, and the product less – the usual problem in a recession.

The retreat of the ice sheet allowed animals and their hunters to move into Scandinavia, but it took time. There were hunters in southern and central Sweden by about 5,000 years BC (7,000 years ago), not long after the final flooding of Doggerland, and at about the time the mammoths became extinct in Russia. These hunters had clearly followed their prey animals into the north.

Elsewhere, if the hunters stayed put, greater reliance now had to be placed on collecting vegetable foods, though the advantage here was that these food resources became much more available now that the land was warmer. In neither case therefore were the hunters and gatherers in continental Europe seriously disturbed in their ways of life, either by moving with the animals, or by staying put, though since it tended to be women who were the gatherers and men the hunters, a certain alteration in power within the families may have occurred. Hunters of mammoths continue to hunt mammoths, if in a different part of the world; hunters without mammoths to prey on continued to hunt other animals in a new part of the world, though they had to adapt their work to a more difficult environment. This was not the case, however, in the North Sea lands.

Moving to Norway

The problem in Doggerland was that the land was flooded, probably over a relatively short period. At the same time in the nearby lands, which are now England and the Low Countries, forests became rooted and developed, slowly spreading northwards with the increasing temperature – this probably did not happen in Doggerland, since the flooding was so quickly accomplished. Fishing has been identified as one of the responses to the change by the people of the area, and this had no doubt been a practice wherever the hunters had access to the sea or to a lake even in the Ice Age. And investigations recently in Norway have added a further dimension to this matter (Fig. 18).

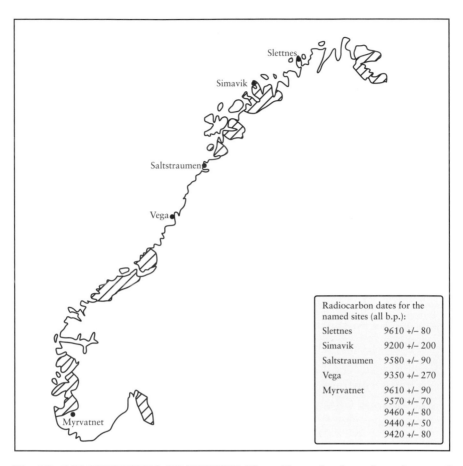

Fig. 18. COLONISATION OF NORWAY. The evidence for the early settlement of people in Norway is mainly from the radiocarbon dates of a series of Mesolithic sites which cluster around 9,500 years ago. The areas of these 'Pre-boreal' sites are shaded. (Some other areas were probably also settled, but the evidence has been destroyed by a later sea-surge.) Note that the dates for the southernmost and northernmost sites are almost identical. Slettnes and Alta Fjord are rich in rock carvings only slightly later than these dates. The coastline had been free of ice for only a few centuries when it was occupied by humans.

The whole coastal lands of Norway, from the south-west opposite Scotland as far as the North Cape, had been covered by the ice, but the warming allowed the warmer sea to begin the melting of ice all along the Norwegian coast, as far as the North Cape. These coastlands were colonised by fishermen and hunters at more or less the same time. Within only a few decades, the whole of the coast – about 1,000 miles long, not

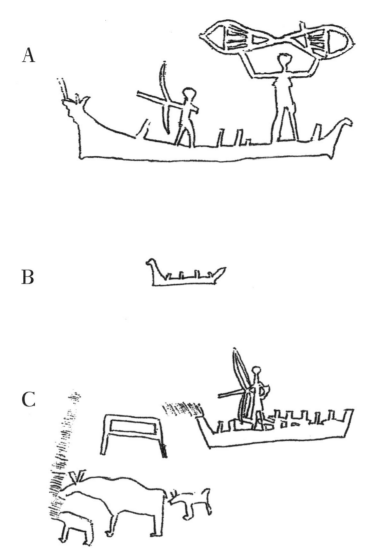

Fig. 19. EARLY NORWEGIAN SHIPS. Northern Norway was settled over 9,000 years ago. In the next thousand years or so the people chiselled pictures of animals and ships on the local rocks, particularly at Slettnes and in Alta Fjord. The animals include bears, reindeer, elk, and dogs – the animals to be expected in the far north. The ships appear to be made of bone and wood frameworks covered by hides. One (A) seems to be about six metres long, judging by the scale of the standing man. It is being used for hunting and fishing – one man holds a bow, the other a net.

All the boats are shown with an elk's head on the bow, either as a hunting disguise or as a sign of the materials used (bone, hides) in its construction. Several of the drawings show little detail, but A and C are both hunting scenes. The people must have arrived in the north by ship, and this may be one reason they depicted them in their rock art.

including the fjords – was occupied by humans for the first time (Fig. 15). Not long before the colonists arrived, this had been an icebound coast, with the huge ice sheets spreading from the mountains and into the ocean as pack ice, though there had always been a stretch of open water not far from the coast, as the Ice Age version of the Gulf Stream had its ameliorating effect; there was also a long narrow bay curling round the southeast and south Norwegian coast, separating Norway from Doggerland.

Most of Scandinavia was still under the ice as the colonisers arrived, so the available land was not extensive. Further, the period of settlement is described as the 'pre-boreal', that is, before the trees which were advancing northwards through central and Western Europe had actually reached Norway; it was therefore tundra. Radiocarbon dates of the earliest settlements concentrate around 9,500 years ago, the earliest in the south near Myrvatnet near Stavanger gives a date of 9,610+/-90 b.p. ('before the present'), and in the far north, at Slettnes, one of 9,610+/-80 b.p. – effectively these were simultaneous, and indicate that settlement from the south to the north took place at virtually the same time.

In the circumstances of the land's condition it is hardly surprising to find that the colonists were basically fishermen. They lived by sea fishing, and by hunting what could be caught, though there cannot have been much animal prey available until a more decisive retreat of the ice sheet.

The colonists left evidence of their lives in rock carvings and drawings, which have been found particularly in northern Norway (Fig. 19). These drawings, pecked and chipped into the rock faces, are very difficult to date, and some or all of these northern examples are no doubt much later than the original settlement of the area – the first settlers must have had more urgent tasks than art – but they do indicate the animals which were being hunted. They show that they hunted elk, a tundra-feeding animal, and the archaeological remains indicate that they also hunted seals, and fished the sea. The pictures also show that these people had seagoing boats, which is what one would expect.

Ice–Age Europe was not a land of boats; the earliest known boats date from the period when the ice was melting and retreating, and the forests were advancing. The reason is obvious. Boats are usually made out of tree trunks, stripped of their bark and hollowed out, and they were used to sail along the rivers, perhaps even as logs, without much alteration. One of these dugout canoes, from western Germany, is dated to about 7,000 years ago, which is round about the time when one could expect trees of a sufficient size to have grown and become available for hollowing out into dugouts. The arboreal colonisation of Europe happened in stages, with birch and hazel usually among the first to take over areas of tundra, followed by pines, and then, but not until about 8,000 years ago, mixed-oak forest. Oscillations in the climate, with returns to cold conditions at times, on several occasions forced a retreat of this vegetation (Fig. 20). And, of course, only the bigger – and therefore older – trees were of much use for boats. But it is noticeable that just as soon as suitable trees were available, men used them to make boats.

On the other hand, this is only the physical evidence, the discovery of the remains of actual dugouts. It is clear from archaeological remains in settlements – Star Carr is an example – that boats of some kind were used in lakes and rivers, and at sea, before the dated physical remains of boats. Since the people consumed fish, lived on islands, and travelled across rivers and straits, it follows they had boats. So it is probable that some sort of watercraft – probably boats made of a wooden or bone framework covered by hides, as used later in Ireland – existed before the earliest archaeological evidence which survives, that is, before 7,000 years ago; and Star Carr is 2,000 years earlier than that.

The Norwegian evidence comes from a time when there was nothing in the way of tree cover of any size in that country – indeed, northern Norway, where the archaeological evidence for boats, fishing and the settlements is most comprehensive, has few trees of any size even now. This indicates that a different type of craft was used along the North Atlantic shores, ones not made of wood – again skin boats are likely. The new land in Norway was still largely covered by ice; indeed, the highest mountains in Norway are still covered in snow all year round even now. So at first only the coastal areas were available for human settlement, and even they had few resources. In the north, where some of the best of the ancient rock carving pictures are preserved, the trees tend even now to be small and stunted and slow-growing, with very little wood in them; they are certainly not the sort of trees with which to make a boat of the dugout type. This was the normal

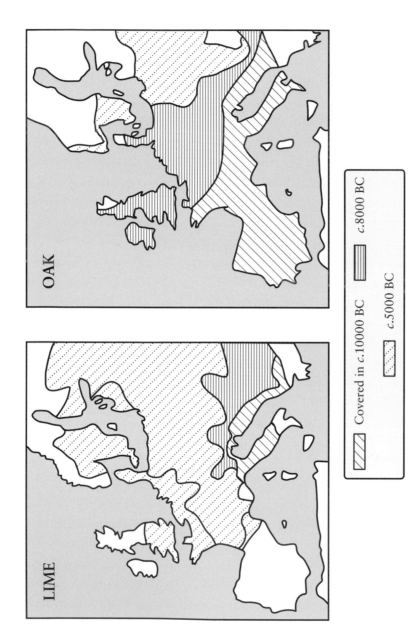

Fig. 20. ARBOREAL COLONISATION OF EUROPE. The maps show the area where lime and oak grew at three stages in the post glacial warming in Europe. It took 5,000 years for them both to reach their maximum extent (reached by about 5000 years ago). Note that lime trees have receded from their southern area as well as colonising northwards. Other trees show similar expansions, birch and hazel moving in advance of all the rest.

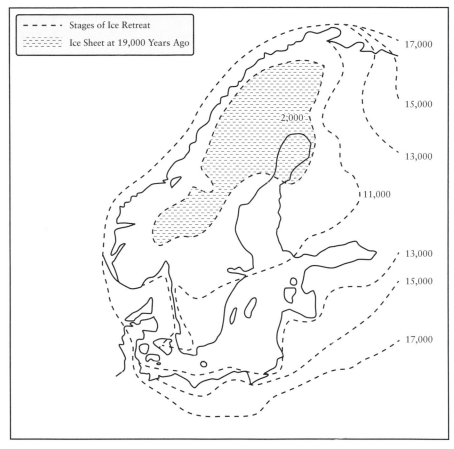

Legend:
- - - - - Stages of Ice Retreat
:::::::: Ice Sheet at 19,000 Years Ago

17,000
15,000
2,000
13,000
11,000
13,000
15,000
17,000

Fig. 21. THE RETREAT OF THE ICE. Even before such sophisticated methods as radiocarbon and AMS dating the stages of the retreat of the ice in Scandinavia had been worked out by varve analysis – counting the layers of silt deposited in lakes in Sweden. From its greatest expansion at about 17,000 years ago, the ice melted, and by 9,000 years ago, largely vanished. It will be seen that it was only by about 10,000 years ago that the Norwegian coast was free of ice, and so available for human (and animal) settlement.

condition of coastal Norway even for some time after the ice melted – the 'pre-boreal' is indeed an accurate name for it. Oaks, for instance, while not the earliest trees to survive, scarcely reached any part of Britain before 9,000 years ago, and reached no part of Scandinavia for another 2,000 years after that. The strip of coast of Norway between the sea and the ice was the only available land at the time – in the sense that it was not ice-covered, and so at first had no preliminary tree cover – from between about 10,000 and 9,000 years ago, which is just the time when the land was settled. And yet the new settlers can only have arrived by sea, since there was no land connection between the Norwegian coastlands and anywhere else at the time. So the boats that were thereby being used, both for fishing and for reaching the new land, cannot have been made of wood, but rather skins, for there were no other materials available.

It is worth taking a moment to consider the scale of all this. The Norwegian coast stretches from the southern point near Kristiansand to Varanger Fjord in the north for well over 1,500 kilometres; it is pretty inhospitable even now, and much of the coast is cliff-bound; there are many islands off the coast, though these are often small and low, and would scarcely support very much in the way of animal life for hunters to prey upon, though fishing would be possible. There are also the fjords, but the further inland one went, the steeper the enclosing lands, and the closer one came to the ice which covered the interior – and glaciers will have blocked the inland ends so that icebergs, or at least large lumps of floating ice, would often occupy the fjord as well. The land is exposed to a high rainfall, and to frequent Atlantic storms. And, of course, the ice was still nearby at 10,000 – 9,000 years ago. Yet over a period of only a few decades the whole of this coast, as far as the North Cape and beyond, was occupied by a newly arrived population.

The colonists can only have arrived by boat; there was simply no other way. As the ice retreated from the coast and into the Scandinavian interior, the northern North Sea became open water all the way between the coasts of England, Doggerland, and the Netherlands. North of that area was the deep trench called the Norwegian Deep, encircling Norway's southern coasts from the Skagerrak to the Atlantic; it was never dry, and became open sea as soon as the pack ice melted. It follows that the colonists could only reach Norway by boat, and that they were therefore already familiar with boats and with the sea, even before they set out. This rules out people from continental Europe, where there were as yet no boats, nor any use for them, and when boats were first constructed there they were made out of wood, and indeed it also rules out anyone south of the North Sea as far as the English Channel, for these coasts were only formed about 7,000 years ago, long after the Norwegian colonisation.

The obvious source of this migration is therefore Doggerland, the land which was even then disappearing under the North Sea as it became flooded. These were people who had already come under pressure as the land became steadily more waterlogged. The great rivers of the disappearing land had already been providing an alternative diet of fish, as did the flooding North Sea. The inhabitants had no doubt slowly changed their diet to one in which fish was emphasized, as hunting by land became more difficult, and so meat became less important in their diet. The North Sea lands had become less hospitable to the larger animals, and the people had evidently therefore developed an ability to build seagoing craft, no doubt beginning with river voyages. They had the source materials for the boats, in the bones of the large animals they hunted and hides from those same animals – they had used both in making their tents, and some would see the boats as merely upside-down tents, at least at first.

Their land was disappearing at such a rate that they were compelled to leave. Some no doubt retreated before the advancing waters into England and Scotland, or into the continent anywhere from Denmark to France. But some, perhaps because they were cut off by the advancing waters, and certainly because they had the use of boats which could accomplish the voyage, went north to the new land of Norway which had recently emerged from the ice.

The voyage was fairly short, but by no means easy. It involved at least one substantial open-sea voyage, across the northern North Sea and over the Norwegian Deep. Then there was the voyage along the intricate island-rimmed Norwegian coast, exploring the fjords, searching for places to settle. The available places for settlement were small and relatively few; only small groups of people would be able to settle in any one place. Some of the earliest voyagers will have sailed all the way as far as the north of Norway, which they reached as early as other voyagers arrived at anywhere else in the country. The land in the north has more extensive lowland than that in the south, and might well have been more familiar and welcoming to people from the low-lying Doggerland area. It cannot have been much colder in winter; the Gulf Stream's effects made the climate much the same along the whole length of the coast.

The process of settlement went on for several decades, perhaps for up to a century, judging by the radiocarbon dates (though these are only approximate, and cannot be pressed for pinpoint accuracy). We have to assume that, as with the settlement of Australia thousands of years before, prospectors went out first to explore the new land, and that they were then followed by a steady trickle of families; this would explain the near simultaneity of occupation in the southern and the northernmost regions. There must have been preliminary voyages even before the prospectors went out to select suitable settlement sites, since they obviously knew of it from the start of the migration. It clearly assumes a preliminary knowledge of the existence of Norway and of its coast, perhaps gained by fishermen; but much later Irish curragh-voyages are known to have reached Iceland, which is an even greater distance.

The implication of all this is that the use of the North Sea waters for voyaging and fishing had been quite extensive before the settlers set out. The whole settlement enterprise was surely an organised communal effort. The prospectors would need to be financed in some way with boats and supplies, and the migrating families would need at least several weeks' supply of food for the voyage, and more still for the first months of their new settlement, until they found out how they could exploit the local resources of the new land. Those who settled first presumably helped those who arrived later, just as those who were the first settlers had been assisted by those they left behind.

We can obviously appreciate the difficulties these people faced, and praise them for their ingenuity, their seafaring ability, and their determination. The process must have involved long discussions, communal decision-making, exploration in advance, detailed planning, storage and accumulation of food, and the advance selection of the new homes, and will have undoubtedly involved considerable cost and probably a fair number of casualties. Some of the people in Doggerland probably refused to move and stayed in the North Sea lands, either to follow on eventually or to drown (though the island which eventually was sunk to become the Dogger Bank did continue to be above the waves for several thousand more years). Some of those displaced will have gone elsewhere, into England, or into the continent. But it is quite probable that the majority of the people in Doggerland chose to move northwards. They are the ancestors of the present Norwegian population – and of the Norse Vikings of the Middle Ages, who made the return voyage. (Some Norwegians have an adaptation to the cold, enabling them to work effectively where others cannot; this may be an inheritance from the Ice Age.)

The Conservatism of the Survivors

Let us, however, not be carried away by contemplating the difficulties the former Doggerlanders faced, or be too admiring of their achievement, though we can surely appreciate both qualities. These people were in fact being extremely conservative, in the same way as the mammoth hunters who followed their prey northwards and eastwards. Their purpose was simply to change the place they lived in, so that they could continue the same way of life they had grown up with. Their solution to the problem which was presented to them by the ending of the Ice Age was to continue to live close to the ice, as they always had. In that sense they were fortunate in where they lived, since the development of the Gulf Stream of warmer water flowing from the south assisted in pushing back the edge of the ice along the Norwegian coast, therefore providing them with the long inletted and islanded coastlands very early in the process of the ending of the Ice Age. Without this there would have been no possible new home for them to go to and they would have been compelled either to die out or possibly move south or east. Therefore they took advantage of their unusual geographical situation to continue living the same life that they had developed during the Ice Age itself. In this, of course, they were adopting the same strategy for coping with the climatic change, with the global warming of the end of the Ice Age, as the mammoth hunters had, who had also simply relocated themselves into more northern and eastern lands which were still tundra and permafrost, in order that they should continue their same way of life.

It is worth noting here that the people who lived on in the warmer climate of the southern part of Europe continued to live much as had their ancestors, though no doubt thankful for the absence of ice and long winters and for the increase in temperature. They would seem to have moved out of the caves which sheltered those ancestors, for the open air was now clearly preferable to the cool caves. In some areas, however, caves and rock shelters continued to be used, and in Iberia there are records of their life during this archaeological period, which is called the Mesolithic, and the inhabitants continued the practice of painting the walls of the caves. A group of these pictures are shown in Gallery III.

Beringia

The North Sea difficulties for the people thus displaced by the flooding of Doggerland are paralleled elsewhere in other parts of the world. The varying responses of other groups will be considered in subsequent chapters. In one area, however, the resemblance to what happened in westernmost Europe was very close. This was at the further end of the Eurasian continent, in the land which has been given the name 'Beringia' (Fig. 22).

This is named, of course, from the Bering Strait, the shallow sea between Eastern Siberia and Alaska which was dry when the sea level was lowered by the formation of the ice caps. Like the North Sea, it was a plain covered in tundra and 'mammoth steppe', an attractive area for mammoths and other Ice Age herbivores. It was also effectively an isolated land. To the east was the great North American ice sheet, centred partly on the Rocky Mountains and partly on Greenland and northern Canada: in combination these two huge ice sheets blocked the way into the rest of America. To the west, in the modern Russian Far East, there were negotiable passages for animals and hunters between the ice-covered mountain ranges, though they were fairly few and both difficult and narrow. In effect Beringia was a subcontinent 3,000 kilometres in size from east to west, and half that from north to south, including Chukotka (the eastern end of Siberia), the dry Bering Strait, and Alaska.

Since there were game animals to hunt, it is not surprising that men lived in the area, having arrived towards the end of the Ice Age, but long after the arrival of other groups had reached Europe and China (that is, round about 14,000 years ago, 15,000 years after China had been populated by hunters). They lived like everybody else at the time, by hunting and gathering, and like all the hunters of the Ice Age, they moved with the seasons and the animals, leaving evidence of their lives in the form of the remains of their camp sites, which were usually placed close to lakes and rivers. In other words, they were living more or less the same life as the people of the caves and camp sites of Europe; close parallels would be with Meiendorf and the hunters of the Ukraine. It was only when the ice began to melt that this land became worth settling. That is, the people moved into Beringia/Siberia/Alaska in order to continue

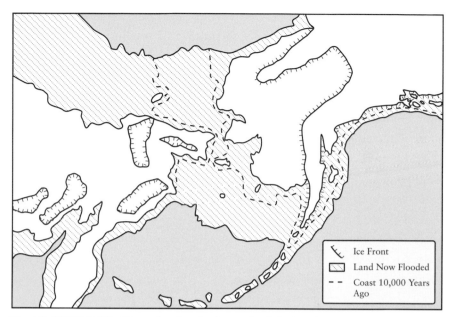

Fig. 22. BERINGIA. The reduction in the sea level during the Last Glacial Maximum exposed large stretches of land between Alaska and Siberia. The great ice sheets and the cold desert of Siberia effectively isolated 'Beringia' from the rest of the world. At 3000 miles across it was a virtual continent in itself. The return of the sea, however, severed the two main parts by about 10,000 years ago, though travel between them was no doubt still possible.

their old way of life of hunting the larger animals of the cold lands. In this they resembled the people of Doggerland, or those of interior Europe, or the mammoth-hunters of the Ukraine. Here, however, the future was to present them with a different set of opportunities and possibilities. One of these opportunities was to move south into other parts of America, a matter to be dealt with in the next chapter. Here I want to consider those who stayed put.

In Doggerland, as it was being flooded, the option of staying put did not exist, and in Western Europe it existed only by forcing change on the people when the animals they hunted disappeared and the tundra became forest. In Beringia considerable parts of the subcontinent between the Pacific Ocean and the ice survived above the new sea level, though the lowest lands were flooded and the subcontinent became divided into two, with Chukotka to the west and Alaska to the east, both of which regions were largely areas which remained as cold tundra lands. The inhabitants of Beringia therefore did not have the option of staying or moving; they had to continue the same life as before, though it seems probable that the mammoths died out fairly quickly. Further, the climate did not change all that much, being still Arctic, with short summers and long winters, all the more trying than in other, more southerly, areas by the long dark winters. Some may had left by going west into Siberia, but perhaps not many.

Those in Beringia/Alaska therefore found that their territory was steadily reducing in size until about 8,000 years ago, when an open waterway developed separating Alaska and Siberia. The strait was, and remains, fairly narrow, though it widened considerably after the first water breakthrough. Yet at that same time the great ice sheets began to melt, so, just as some of their hunting areas were drowned, other areas became available, though only slowly. The situation is clearly quite congenial enough to persuade part of the hunting population to stay in place there, while others found their way out, either west, or along the coast, or by other means (Fig. 23).

Yet here in the north, as elsewhere, warmer temperatures encouraged the spread of trees. In Alaska the earliest trees were poplars and willows, which would seem to have been already present in hidden sheltered areas in the land – which are called 'refuges' by palaeobotanists. These trees were followed north by spruce and alder. The poplar/willow spread began about 11,000 years ago, earlier than the sundering of the subcontinent, but the alder/spruce group, coming from outside, took much longer to spread, and even now these trees have not occupied the whole of Alaska (though the current episode of global warming will no doubt encourage them to complete the process). By about 5,000 years ago they had more or less reached their present range, but there are still areas of muskeg and tundra intermixed with the spruce forest in the country.

This timing meant that the inhabitants were not under any great pressure to change their way of life, except for the reduction in the numbers of the larger animals, whose products could be replaced by other prey with some extra work. As in Norway, some of the larger animals did survive, with elk and caribou present in considerable numbers. Those people who remained in Alaska became the ancestors of the two modern hunting groups who survive there to this day, the Aleut and the Inuit.

Along the Alaskan Peninsula and the Aleutian Islands, which stretch out to the south-west from the American mainland, the land and islands had been extensively glaciated, and the ice cover probably survived for a long time (as in Norway). The people now called Aleut developed as a maritime population, turning their hunting skills towards the sea's prey, making their living by fishing and collecting shellfish, and by hunting whales and seals.

Their new life was very like that of the earliest Norwegians, though the Alaskans only had to move a short distance to their new homes, from the inland to the shore (once the intervening mountain range had shed enough ice to open routes from the interior to the coast). In a sense they replaced the great land animals, such as the mammoth, by the great sea animals, the whales, or at least by seals, walruses and sea lions. These people became the modern Aleut, skilled fishermen and sailors.

To the north, in the interior of the Alaskan mainland, a second group adapted itself to a life actually on the ice. These became the ancestors of the Inuit (who for a time have also been called Eskimos). They remained hunters, depending largely on hunting seals and similar aquatic mammals, but any land animals which they found on the ice as well. They developed a lifestyle by which they could live actually on the ice, not simply close to it, and were successful enough so that they were able to spread rapidly across the Northlands as far as Greenland. Possibly they were pushed into their radical shift and expansion of location by the northerly spread of the new forests. They exploited the resources which were available to them in the north, and on and under the ice – furs, bone, ivory, skins, fat, even ice and snow – and developed a set of clothing and weapons and tools and shelters which were particularly and peculiarly useful and adapted to their situation – though these expedients were usually only improvements on what had been worn and used already by hunters in the Ice Age. Both Aleut and Inuit, of course, used canoes – kayaks and uniaks – and both remained hunters.

These responses to global warming by the Aleuts and the Inuit was very much the same as that produced by the North Sea Norwegians and the Eurasian mammoth hunters. The Inuit, like them, refused to accept that what was happening compelled them to change their lives, and continued with their old way of life as long as they could, even moving onto the ice

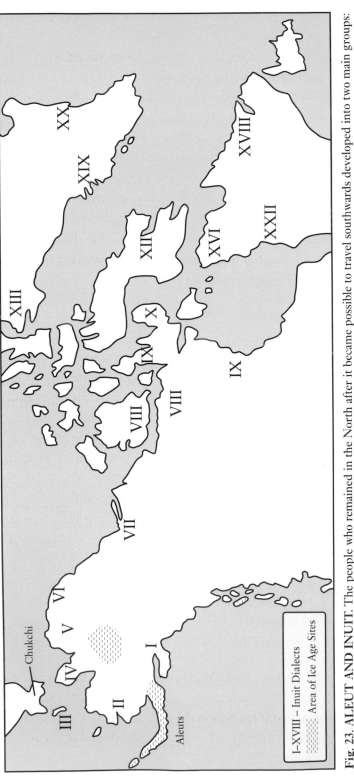

Fig. 23. ALEUT AND INUIT. The people who remained in the North after it became possible to travel southwards developed into two main groups: the fishermen Aleut, who stayed in their original homelands but lived by the sea, and the Inuit (who used to be called Eskimos) who adapted their life to the cold and spread across the far north as far as Greenland, and did so in perhaps no more than 1,000 years. They remained in contact throughout this huge area, and their several dialects are still mutually intelligible. (Other Inuit live in Siberia; they are also related to the Chukchi of Siberia, who have become herdsmen.)

Chukchi

Aleuts

I–XVIII – Inuit Dialects
Area of Ice Age Sites

when the forests restricted their hunting activities. Similarly, the Aleut continued to hunt the great beasts, but learned to do so at sea when their land prey had disappeared, or were drastically reduced. The two peoples were, it seems, originally one, but they separated as the Aleut moved to live near the sea and the Inuit remained inland – their languages are closely related, but have evolved to cope with their differing lives and surroundings.

The mammoth hunters in Eurasia and Beringia therefore had to eventually die out or change, because their chosen prey died out. This was, however, not a final choice forced on them for several thousands of years; the Norwegians, the Aleut, and the Inuit found that they could continue a successful hunting life on the edge of, or actually on, the ice even after the Ice Age itself officially ended.

Conservatism Again

The aim of all these peoples was wholly conservative, in that they attempted, in many cases successfully, to continue to live their original way of life without any change, except in having to move their living space elsewhere as their prey moved – but being migrants already this was not a great hardship, and did not compel them to change their lives. Where caves were available they continued to live in them. Of course, once the period of global warming after the Ice Age ended and the climate stabilised, from about 7,000 years ago – though the ice melting continued for a time after that – both Inuit and Aleut and their cousins in northern Eurasia, had to change, but the change was in the prey they hunted, not in their hunting lifestyle. Yet even these changes came about through outside influences, not because of their need to cope with the change of climate. They have also been tenacious in resisting any more change, and still are, many of them accepting only those elements of other lifestyles which fit into their own. The Norwegians changed more than the others, perhaps, but the Aleut and the Inuit have proved to be most resistant to further changes, their lives having been successful adaptations to particular conditions, and have lasted perhaps up to 10,000 years. They are still, therefore, conservative, a response which has served them well in the past, and which is perhaps bred into them by now by the experience of coping with the end of the Ice Age; this is why they remained in the north.

Further Reading

For the condition of Europe in the Ice Age see any account of the Palaeolithic which is not obsessed with stone tools, such as Graham Clarke, *The Stone Age Hunters*, London 1967, which, though now rather old, is concise and well-illustrated, or T.G.E. Powell, *Prehistoric Art*, London 1966, and more recently the first chapters of the *Oxford Illustrated History* noted in Chapter 2. For North America see E.C. Pielou, *After the Ice Age; the Return of Life to Glaciated North America*, Chicago 1991, though this is concerned with animals and plants rather than mankind. The archaeology of the mammoth hunters of Europe is also discussed by Richard G. Klein, *Ice Age Hunters of the Ukraine*, Chicago 1973. Brian Sykes, *The Seven Daughters of Eve*, London 2001, has expounded his genetic theory of the small original population of Palaeolithic Europe (he has now expanded the number of the original European women to twelve or so).

For the Norwegians, see M.A.P. Renouf, *Prehistoric Hunter Fishers of the Varangarfjord, North-eastern Norway, Reconstruction of Settlement and Subsistence during the Younger Stone Age*, British Archaeological reports S487, Oxford 1989, and H. B. Bjerch, 'The North Sea Context and the Pioneer settlement of Norway', in A. Fischer (ed.), *Man and the Sea in the Mesolithic*, Oxford 1995; there are also articles on Norway in the papers of the Edinburgh symposium (see Chapter 2) and in a later meeting in Belgium: S.E. Nygaard, 'Mesolithic Western Norway' and S. Bang–Anderson, 'The Myrvatn group, a Pre-boreal Find-complex in south-west Norway', in F.M. Vermeersch and P. van Peer (eds.), *Contributions to the Mesolithic in Europe*, Leuven 1990. The rock art showing ships is discussed by S. Stolting, 'The Boats of Slettnes: Sources of Stone Age Navigation', *International Journal of Nautical Archaeology*, 20, 1997, 82–193. A discussion of their putative original home is by A. Gob, 'The Early Postglacial Occupation of the Southern Part of the North Sea Basin', in the CBA research report noted in Chapter 2. This is now overtaken by the new work on the North Sea and Doggerland, as in V. Gaffney, Kenneth Thompson, and Simon Fitch, *Mapping Doggerland*, British Archaeological Reports, Oxford 2007, and in V. Gaffney, S. Fitch, and D. Smith, *Europe's Lost World, the Rediscovery of Doggerland*, Council for British Archaeology Research Report 160, York 2009, which includes short sections on Beringia and Sahul. For Sweden, see Mats Larsson, Geoffrey Lemdahl, and Kerstin Liden, *Paths Towards a New World, Neolithic Sweden*, Oxford 2014.

For the development of the Aleuts and the Inuit see the brief discussion in Brian M. Fagan, *People of the World*, New York 1986. An agreeable account of the search for Beringia is in the biography of its 'discoverer', David M. Hopkins, by Dan O'Neill, *The Last Giant of Beringia, the Mystery of the Bering Land Bridge*, 2004; Hopkins' own account is *The Paleoecology of Beringia*, New York 1992.

Gallery III

Scenes of Mesolithic Life in Iberia

Caves in Spain have just as good paintings as anywhere else. Indeed, the cave at Altamira was one of the first to be studied carefully. The range of pictures is also wider, and the practice lasted longer than in France, well into the Mesolithic. Hunting, however, remains the prime activity, as in this depiction of a hunt from the Loc Caballos Cave, with deer being driven towards a line of archers.

III.1 Loc Caballos Cave – deer hunt.

One of the human activities which developed in the Mesolithic age was organised warfare. Picture III.2 is part of a larger depiction of battle between two 'armies', together amounting to perhaps thirty men, all archers, who have come to very close combat – they are thus actually hunters, who needed to get close to their prey to ensure success. An example of gathering is the person (a woman?) taking honey from a hive, apparently without fear, despite the clouds of bees all around him/her.

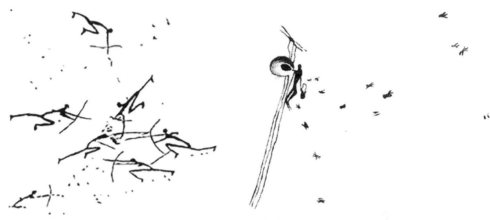

III.2 Fragment of picture of a battle.   III.3 Stealing honey.

One of the distinctive elements in the Spanish pictures is the variety of subjects (as in III.2 and 3), and in the variety of techniques used by the artists. The stick figures in III.4 are crude and primitive, and very difficult to interpret, but some at least have hats on whilst others are sitting down and altogether the scene is clearly a crowd of people.

III.4 Stick figures.

The two animal pictures (III.5) show another technique which is also used in the French caves of simply outlining an animal and surrendering it as a representative of the hunted species – a mammoth, a rhinoceros. And others rather more faint – all drawn without hesitation or alteration.

III.5 Prey.

Chapter 3

Escaping the Ice

Once a way of life is established and is successful, as was that of the hunters of Ice Age Europe, it takes a great effort, or a great catastrophe, or perhaps great imagination, to decide to change it; and then there is needed great and prolonged determination to achieve the change. It is clearly easier to avoid such decisions and disruptions and to settle for continuing the life one is already living, even if moving elsewhere is necessary.

The hunters discussed in the previous chapter adapted themselves to the end of the Ice Age by merely moving their homelands so as to maintain their livelihoods, or not moving if they were able to continue as before with only a minor adaptation, such as substituting different animals for their preferred prey. Eventually, of course, they had to adapt to still further changes, the mammoth hunters to the disappearance of the mammoth, the Inuit to the advance of the forests, and life on, rather than close to, the ice, but in all these cases the main aim was to change as little as possible, and for a long time, for several thousands of years, they both succeeded; the Aleut seem to have changed more, but in effect they continued their hunting life, but for a different prey. Here I want to discuss a different response, one which, like that of the new Norwegians, involved people moving, but in these cases to a land where new opportunities existed.

The incidence of ice sheets across the northern hemisphere was erratic. High mountains, of course, attracted ice caps simply because their height meant the air at the highest parts was cold and thin, and snow, once it had landed, did not melt, and high latitudes attracted ice caps because the radiation from the sun was attenuated and so snow accumulated, again failing to melt. The combination of mountains and a high latitude automatically produced ice, as in Scandinavia, and the several mountain ranges in Siberia, in the Rocky Mountains in North

America, and in the southern Andes in South America. Once established, these ice caps reflected more of the sun's radiation away, and so the caps were perpetuated. Variations both in altitude and latitude meant that in other areas across the north the ice regions alternated with tundra, notably in the eastern Siberian mountain ranges.

Beringia (Fig. 22)

This condition in effect produced wide areas of the north in which tundra predominated, and these were the territories which remained a favourable environment for the larger cold-adapted animals of the north, areas which are sometimes called 'mammoth steppe'. One of these, paradoxically, was Alaska–and–Eastern Siberia, which are almost typical now as northern places of cold and snow. The fall in the sea level linked these two areas by land, as noted in the previous chapter, along with the responses of two groups of inhabitants to the change. The name Beringia for this revealed land is a term which should be used to include the ice-free areas of the nearby Siberian and Alaskan mainland as well, making the partially isolated region about the same size as that part of contemporary Europe which was free of ice at the time, a subcontinent in fact.

This territory was long uninhabited in the worst of the Ice Age, but as the grip of the ice lessened, it was colonised by the cold-climate animals, including mammoths, and by the men who hunted them. The new inhabitants lived the usual life of the Ice Age hunters, camping at comfortable spots, often repeatedly in the same place, retreating to caves for the winters, hunting mammoths, reindeer, elk, and so on, and moving with the migrating herds. The difference from Europe was that they had ice more or less all around them, in the Siberian mountains, if with tundra-clad gaps, and to the south and east, where much of the rest of North America was covered in gigantic ice sheets – as was the long tongue of the Alaskan Peninsula and the Aleutian Islands; to the north was the Arctic Ocean, rimmed at least by pack ice, while more of this pack ice extended partway into the north Pacific Ocean from the southern edge of the land; Beringia was a relatively ice-free island amid the ice caps.

The inhabitants of this isolated region reached it about 13,000 or 12,000 years ago, from the lands to the south and west, the cold steppes of eastern Russia and Mongolia, but were then prevented from moving further on into America by the immense ice barrier which covered Canada. This conclusion is based on the results of archaeological investigations mainly in Alaska, but, as will be seen a little later, it may not be wholly correct. It is a fact, however, that the earliest known (and accepted) human habitation of Alaska dates to about 12,000 years ago or a little later. Some of these were the ancestors of the Aleut and the Inuit, but others chose a different future.

The Ice Age in North America

There were two centres of ice expansion in the north of North America: the Rocky Mountains and Greenland. The latter was much the greater and is usually called the Laurentide ice sheet (Fig. 13). It is estimated that the great dome of ice was up to 2,500 metres thick over Hudson's Bay, and it extended to beyond the modern U.S.-Canadian border – with glaciers extending out even further. The Great Lakes, and the great Canadian lakes, rather like the Baltic, were being formed at this time, partly by the scouring action of this ice, partly by the meltwater at the edges of the ice sheet.

The second area of ice expansion, from the Rocky Mountains, is usually referred to as the Cordillera ice sheet. It reached the Pacific Ocean all along the coast of British Columbia and Alaska, and spread a tongue of ice along the mountains of southern Alaska and the Aleutian Islands, and out from both of these the pack ice covered the nearer ocean. Being on top of a mountain range it was even higher than the Laurentide sheet, though not so big nor so deep. In its spread to the east it met the Laurentide sheet in the (later) prairie lands of Canada, where another series of great lakes – Lakes Winnipeg and Athabasca, the Great Slave Lake, the Great Bear Lake – marks the area of the meeting of the two, and more scraping and meltwater. This meeting region was in fact periodically free of ice as the fronts of the great sheets advanced and retreated with the fluctuations of the temperature. Southwards, the ice covered the highest of the Rockies as far south as southern California.

This immense area of ice obviously imposed a drastic alteration on the geography of the whole continent (Fig. 24), but this was not all.

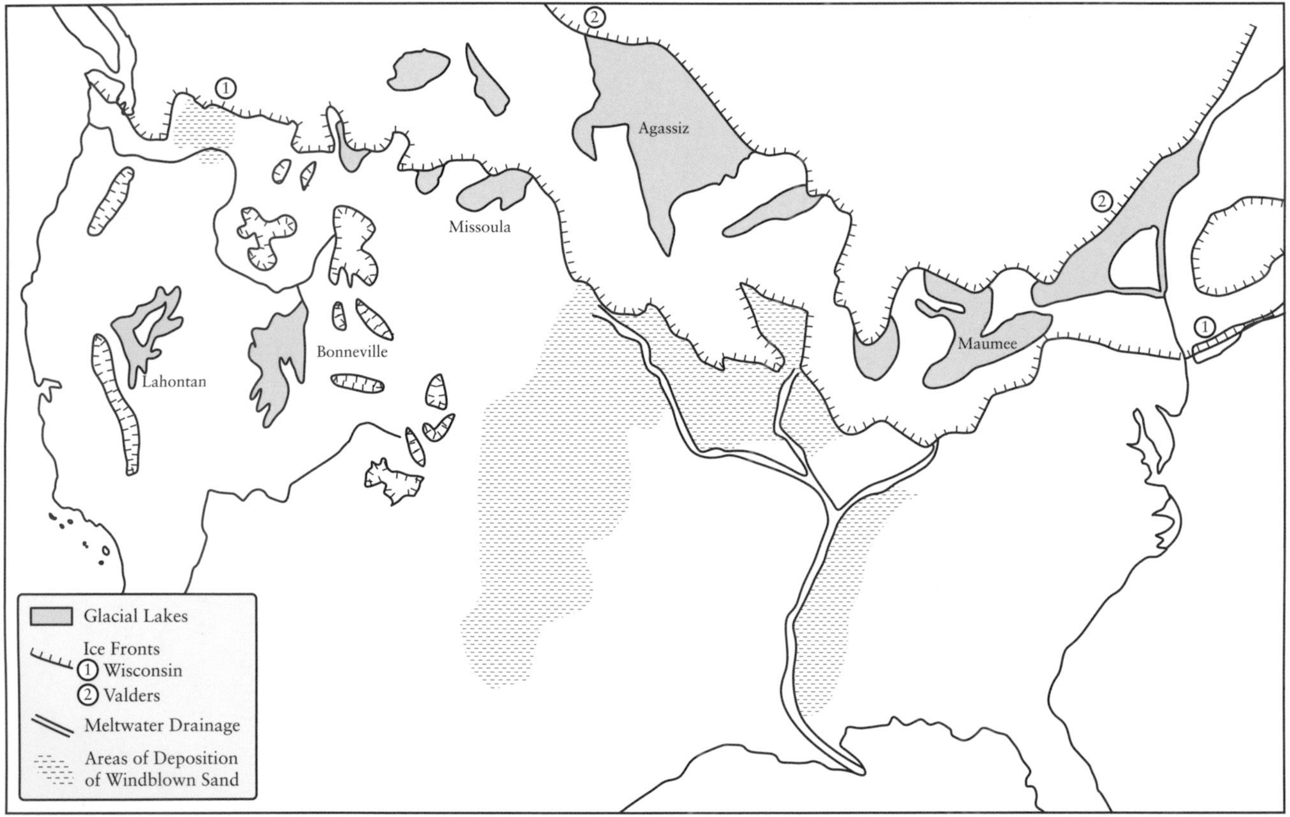

Agassiz
Missoula
Bonneville
Lahontan
Maumee
1
2
Glacial Lakes
Ice Fronts
1 Wisconsin
2 Valders
Meltwater Drainage
Areas of Deposition
of Windblown Sand

Fig. 24. ICE AND WATER IN GLACIAL NORTH AMERICA. The inhospitable nature of Ice Age North America is emphasised by the sheer size of the ice cover, and by its associated unstable lakes and the enormous meltwater rivers. The modern Great Plains were the scene of the deposition of dusty earth, and were very dry. Life was possible, of course, but disaster might threaten at any time.

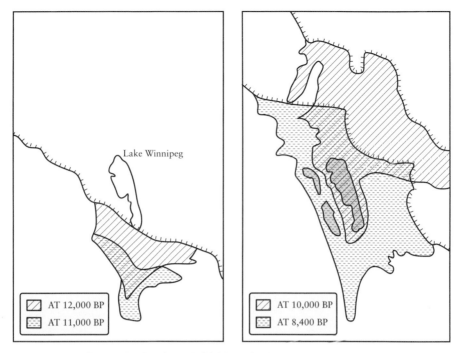

Fig. 25. THE SHIFTING LAKE AGASSIZ. As the Laurentide Ice sheet melted so the quantity of water in Lake Agassiz increased, and its boundaries shifted (always with the ice front on the north-east). Four stages are shown here, the last being just at the time the ice dome over Hudson Bay collapsed and the lake's water drained into the ocean. (The present Lake Winnipeg, a remnant of Agassiz, is shown for orientation.)

The southern edges of the ice sheets, of course, habitually melted every summer, and in North America this action produced a series of large lakes – 'Lake Agassiz', 'Lake Ojibway', 'Lake Missoula', and others – which were in fact temporary in the sense that they have since disappeared (figs. 25, 26, 27). These were the liquid ancestors of the modern Great Lakes. 'Temporary' is an elastic word, and in this case merely means that these Ice Age lakes no longer exist, and that they lasted 'only' a few thousand years before draining away, leaving the present smaller lakes, from Superior to Ontario, behind them, plus the line of the Canadian lakes towards the north. In the Great Basin, between the Rockies and the Sierras, there was another lake, a vastly expanded Great Salt Lake, called 'Lake Bonneville', which was formed not so much by meltwater from the ice as by the increased rainfall of the area; it was landlocked and the

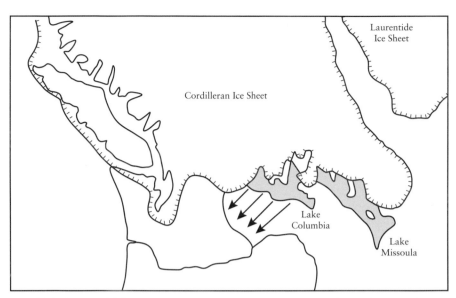

Fig. 26. LAKE MISSOULA. At its greatest extent the ice came as far south as Washington state and Montana. Two glacial lakes were formed, Lakes Missoula and Columbia. They were separated by a lobe of ice which periodically gave way, allowing Missoula to spill over into Columbia which then itself overflowed, sending great quantities of water towards the Columbia River. This happened repeatedly for perhaps 2,000 years, until the ice finally retreated, leaving some small remnant lakes and the clear marks of the repeated floods on the ground, in an area called the Channelled Scablands. Any living being in those areas at the time was clearly more likely than not to be killed by one of these floods – and the Columbia River downstream would clearly overflow its banks.

Great Salt Lake of Utah is its last remnant, becoming as it is now by the solution and concentration of salt by the larger, older lake's evaporation.

Fresh Water Pulses

The water penned up in these lakes had to escape; it was constantly increased and renewed, but was blocked from flowing out in most directions by piled-up moraines and ice. Some of it flowed south as an Ice Age Mississippi River, which was fed by Lake Agassiz, and therefore ultimately by the melting of the edge of the ice sheet (Fig. 25). This great lake occupied a varying area centred on the present Lake Winnipeg. To the west the waters of Lake Missoula drained away along the valley of the present Columbia River, but it did so intermittently and at relatively short intervals as an ice dam repeatedly formed, blocked the flow, broke, and then reformed, and so alternately blocked and permitted the river's flow. More water flowed to the eastward from Lake Ojibway, forming the valleys which now accommodate the St Lawrence and Hudson Rivers. These huge ice age lakes were not only formed by meltwater, but they were also held in place by the great ice sheets, which blocked their drainage routes (like the water of the southern North Sea Lake, similarly 'temporary', and lasting for 'only' a few thousand years). Periodically these blockages opened, either by the ice melting, or in the case of Lake Bonneville, by the collapse of a natural (that is, rock) dam. These were sudden events, producing huge downriver floods, and are called 'meltwater pulses' by the glaciologists. Lake Missoula flooded out along the Columbia River more than forty times over a period of 1,500 years (Fig. 26); Lake Bonneville's flood passed along the Snake River in one huge flood; both of these floods caused great alterations in the landscape – Missoula's floods repeatedly scoured large areas clear of plants and soil. Lakes Agassiz and Ojibway, which eventually grew to such a size that they joined, were the biggest of all these lakes, and produced the most massive floods.

These lakes were huge in extent, holding vast quantities of fresh water, bigger in most cases than all the present Great Lakes put together, but they were constantly changing their sizes and shapes and boundaries as ice melted and froze, and as the drainage channels opened and closed. They lay all along the southern edge of the ice sheets right across North America, and many more existed, always temporarily, than have been here discussed. Most of them were held in place by ice dams or deposited moraines, which are both, by definition, relatively weak structures, and therefore 'temporary'; when these broke great quantities of cold fresh water emptied out rapidly and spilled into the oceans.

The drainings of Lakes Bonneville and Missoula must have been spectacular events but they were essentially local in their effects. On the other hand, when the dams containing Lakes Agassiz and Ojibway broke the result was a flood which had a worldwide effect. This flooding happened more than once, and in each case a great surge of water emptied into the Atlantic Ocean. This happened much less often than with the western lakes, but the quantity of water involved was as a result so great that the whole planet was affected. The ice, and its associated moraines, acted only as a temporary dam holding in the lakes; as it retreated so the ice dams vanished. When that happened with Agassiz/Ojibway, the great rush of cold fresh water fell suddenly into the Atlantic Ocean as these lakes emptied. And it really was sudden. This great meltwater pulse has been adduced as one of the explanations for fluctuations in world temperatures during the period of global warming as the Ice Age ended. The whole of the world was affected by the fall in sea temperature, the reduction in the salt content of the nearer ocean, and possibly by the diversion of the Gulf Stream southwards, all of which combined to cause a temporary revival of the ice advances in both North America and Europe. (The emptying of Lake Bonneville does not seem to have had the same effect, for by emptying into the much larger Pacific the water was more easily absorbed, and anyway much less water was involved.)

The arrival of vast volumes of new water in the oceans brought a rise in the sea level, and this produced a sudden flooding of low-lying land all around the world. It could be one of those sudden rises which finally pushed the Doggerland peoples into moving north to Norway. At least three major meltwater pulses out of North America into the Atlantic have been located, and all of them coincided with major climatic changes.

One of these pulses happened about 12,900 years ago, a second about 11,300 years ago, and a third about 8,200 years ago; these times coincide with the start of the ice re-advances called the 'Younger Dryas', the 'Pre-boreal Oscillation', and the '8.2 KYA (i.e., '8,200 years ago') Event'. This last saw the transfer of an enormous quantity of fresh water into the Atlantic by way of Hudson's Bay and the Hudson Strait, as the ice which was blocking the exit from Hudson's Bay collapsed. It seems also that the dome of ice over Hudson's Bay collapsed at the same time, pushing the waters outwards in a great surge. The combined lakes Agassiz and Ojibway, grown to their joint greatest extent by that time, emptied their waters into the ocean in less than a year. This event altered the ocean circulation in the Labrador Sea (at least in that area, but it is probable that the whole North Atlantic was affected), and this seems to have initiated a renewed cold phase in northern waters – the reduction in the saltiness of the ocean water made it more susceptible to freezing – and, of course, it forced a further rise in the general water level, at first in the north Atlantic, and ultimately in all the oceans.

Scandinavia

None of these changes and 'pulses' was as simple as a mere flood or a tsunami, of course. The possible implications are illustrated by another lake of the 'temporary' type which existed south of the Scandinavian ice sheet, similarly formed from meltwater as the Scandinavian sheet faded. It was first connected to the wider ocean sometime after 11,000 years ago by a channel which opened up across southern Sweden (where there are, as in North America, remnant lakes from this situation even now). In this case it was the rising level of the outside ocean which exerted pressure, so that it was salt water from the ocean which flowed along the channel and entered the 'Baltic Lake'. This gap closed about 9,000 years ago as the ice sheet continued to melt and discharge fresh water, and as the land, relieved of the weight of the ice, rose – the process of 'isostatic recovery' mentioned in Chapter 1 in relation to western Scotland. A new land barrier was formed by the risen land, and a new freshwater lake, the 'Yoldia Sea', then collected behind a combination of the reformed barrier and the continuing, but shrinking, ice sheet; this was reconnected to the ocean about 7,000 years ago through its present connection, the Danish Sound, and this time the current flowed outwards, fuelled by the melting ice, so that the Baltic is now composed of much less salty water that the North Sea. By that time the ice sheet had more or less gone from Scandinavia, reduced to ice on the mountains, so that the coastlines and mountains have been more or less stable since then.

(The North Sea was the scene of similar events, partly caused by the increase in temperature and the weakening grip of the ice; about 9,000 years ago a collapse of land into the sea at Storegga, in Norway, caused a great tsunami which affected all the eastern Scottish coast; other smaller tsunamis about 6,350 years and 2,580 years ago had similar effects. These were much lesser catastrophes than the Hudson's Bay ice collapse – but the Storegga Slide event seems to have cleared all eastern Scotland of human beings for a time.)

It is not always possible to link these events with human history. One is tempted to see the settlements along the Norwegian coast as connected with either the Pre-Boreal Oscillation meltwater pulse in North America or one of the Baltic events – the closing of the channel about 9,000 years ago is suggestively timed, as is the Storegga Slide, though both look to be rather too late for the connection to work, given that the radiocarbon dates from Norway cluster about five centuries earlier. In fact, the effects of most of these northern European events were probably only catastrophic locally; at a further distance the effect was much more diffuse; even with the Hudson's Bay collapse, the change was mainly manifested as a slow rise in the coastal sea level and perhaps a sudden but brief flooding over a flat coastline.

Openings Southwards into North America

At the same time as these events it is clear that, during the time the ice was melting, North America, with its ice, its glacial lakes, and its meltwater catastrophes, remained a very dangerous environment for several thousands of years. The repeated emptying of Lake Missoula every 20 to 60 years, across what is now Montana, Idaho, and Washington states, swept away most of the earth and the vegetation each time – and any animals and people caught in the floods path as well. The enormous Lake Ojibway/Agassiz, two or three times the size of the present Great Lakes, its surface rising to 250 feet above sea level, emptied in no more than a year in the case of the '8.2 KYA Event'. The whole environment of North America changed in that time – a catastrophic event, in truth.

It is also the case that throughout much of this late glacial period, as the ice sheets were contracting, the people who were living and hunting in Beringia had no way out of their sub-continent towards the east and south – into America, that is. Travel westwards into Siberia by land was possible for them until the lowland between Alaska and Chukotka became flooded, and it was also possible even then by boat – both the Aleut and Inuit developed boats. Whether the people of Beringia had boats is a major problem of discussion, and seems unlikely until the Ice Age ended.

It has long been assumed that, as the ice relaxed its grip, the two ice sheets retreated from contact, and became separated, leaving what is known to glaciologists as an 'ice-free corridor' between them. This supposedly provided a long land route to the east of, and parallel to, the Rockies. It would be of varying width, reaching from northern Alaska as far south as Montana and North Dakota (right into the area devastated by the irregular pulses from Lake Missoula).

It has been generally assumed in the past that this ice-free corridor was the route by which the human colonisation of the Americas took place, and on the map, particularly if one uses a small scale map, this looks feasible (Fig. 24). It was, however, a difficult route to follow, not least because it contained a series of lakes, which would block the way. However, recent research now seems to show that some humans already inhabited what Alaskans call the 'Lower Forty-eight' and Central and South America during the Ice Age and possibly even before, though during the Ice Age itself it seems highly unlikely that conditions in most of North America were conducive to any sort of human habitation.

Earlier Inhabitants

Yet the Ice Age, as shown already in the last two chapters, was not a time of continuous cold, but one in which the size of the ice caps fluctuated and the temperature varied. It is clear from the dates acquired from increasing numbers of excavations, that people did live south of the ice in America (meaning the land from the Canadian border to Tierra del Fuego) during the Last Glacial Maximum and earlier (figs. 29 and 30). The two decisive sites are Monte Verde in Chile, and Meadowcroft Shelter in Pennsylvania, both of which have produced material reliably dated to more than 12,500 years ago, well before the original suggested date for the arrival of humans south of the ice. Since it seems clear that the ice was impassable for some thousands of years, certainly during the extreme conditions of the Last Glacial Maximum – though there may have been lengthy periods within the Ice Age when access from Alaska was possible – it is best to assume that the people who inhabited these places were descended from earlier arrivals, though one would be pleased to see rather more evidence for other occupation in the early long period.

It has been suggested that the inhabitants of Alaska 14,000–12,000 years ago did not actually hunt mammoths, but scavenged the bones and tusks of mammoth skeletons – that is, at an earlier period, the area of Beringia was inhabited by these and other cold-climate animals before humans arrived, and so it would have been congenial territory for the people who hunted them. There is, in fact, no archaeological evidence for the presence of humans in Beringia at the earlier period when mammoths were present, though if it is accepted that people reached the lands further south before then, there seems no other route by which they could have arrived.

The argument about the timing of the arrival of humans in the south still goes on among archaeologists in North America. Meanwhile more and more sites are revealed to be older than predicted by the old theory of the arrival after the Ice Age. Discoveries have particularly taken place in South America, but there are also some in North America as well. This is where the unstable climatic conditions in North America during the Last Glacial Maximum are relevant. Ice, floods, lakes, and frequent changes in all these, were all inimical to any stable life in North America as the Ice Age waned. It seems probable therefore that the pre–Ice Age and Ice Age populations were only able to flourish in the south of the double continent – Central and South America, that is – and that any groups reaching the area now forming the United States would move out as rapidly as possible. In addition, the climatic instability in the north was liable to destroy any archaeological remains. Other disciplines have now weighed into the discussion, and both DNA testing and linguistics have produced theories which imply earlier arrivals than the old theory expected. The ice-free corridor therefore has lost its significance as the route south to a large extent, though it does not wholly invalidate the theory of such a movement taking place as the ice retreated. A major scholarly controversy well mixed with obduracy and prejudice is ongoing.

The Ice-free Corridor (Fig. 27)

We have also to contend with the mind-set which assumes that the purpose of immigrants to the American continent was to reach and inhabit the land which became the United States. Given the geographical situation in and after the Ice Age in that area, with floods, ice, and widespread tundra, this territory was surely one of the least inviting places on earth at the time. But the assumption behind the theories of researchers has often been that, since the future United States was the inevitable target of the post-Ice Age migrants, so the 'ice-free corridor' took the place of the later migrant traffic across the North Atlantic. It has taken much argument and much scholarly shaking of heads to provide evidence of earlier inhabitants south of the ice, and even more to get that evidence accepted.

The exercise of a modicum of imagination, however, would have suggested that the ice-free corridor was hardly a highway along which large numbers of people could have moved. First, the population of Beringia was always small and thinly spread – as is that of all tundra lands at any time, and particularly in the Ice Age. Second, the 'corridor' was flanked on either side by the great cold masses of the two ice sheets. It may well have been technically free of ice for a month or so in the summer, but the closeness of the ice sheets will have made it a permafrost region, littered with boulders, large lakes, fierce rivers whose waters were ice-cold, bogs and muskeg, an area producing very little food for the animals which the people hunted, and so very few animals. It was over 3,000 kilometres long, from Alaska to Dakota, and it seems unlikely that there was ever very much in the way of forage for any animals until long after the ice had definitively retreated from the corridor to a comfortable distance and for a lengthy period of time.

Therefore, the corridor could have supported very few of the animals on which any humans who ventured into it could prey. Above all, it will have been frozen every winter, deeply covered in snow and wholly inhospitable. It was much less attractive as a hunting ground than even the shrunken Beringia, and for a long time there can have been little to attract people into it. Nor was there any likelihood that Beringians could have known if there was useful land past the ice. The expenditure of effort involved in transporting a family through the 'corridor', if and when it existed, would be enormous, far more than that required to settle Norway. Further, there are no early archaeological sites in the area of the putative corridor, though this is hardly decisive given the destructive power of the ice. (And it is necessary, I repeat, to avoid the idea that people in Beringia would wish to make such a journey: they were not 'going to' the 'United States'.)

More recent calculations have theorized that the corridor did not actually exist in any meaningful way until the decisive thinning of the ice from about 11,000 years ago onwards – and even when that began, the corridor was hardly available for a long time. And by that time there is already clear and decisive evidence of the existence of several occupation sites south of the ice, Meadowcroft in Pennsylvania (about 14,250 years ago), of course, also Cactus Hill and Saltville in Virginia, and at Hebior

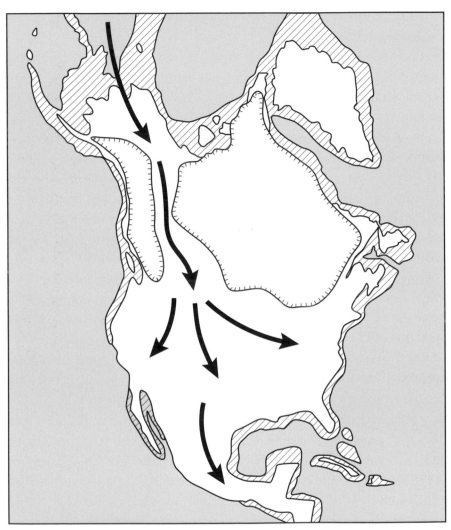

Fig. 27. THE ICE-FREE CORRIDOR. By indicating the corridor in the same way as the southern lands, and by drawing bold arrows (and showing the corridor to be fairly wide) the ease of the movement from Alaska to the south is easily exaggerated. In reality, the corridor was narrow, cold, wet, and very short of sustenance for either animals or men. Even when it was 'open', for 1,000 years it remained unusable for these reasons. Moreover, this version does imply that the coasts were free of ice, and with the lower sea level the continental shelf was exposed, and so a passage south by water was possible.

and Schaefer in Wisconsin, where there are remains which are clearly earlier than the original remains identified as the first arrivals – the Clovis culture. These show that there was human occupation and activity well before the retreat of the ice. And there are other sites further south in

North America. In South America there are even earlier sites: in Colombia, for example (Taima Taima), and even in Chile (Monte Verde), which have both produced convincing early dates and well-authenticated marks of human activity. The Monte Verde site in particular has produced several dates of about 12,500 years ago, a good millennium and more before any definitive opening of the 'ice-free corridor'.

So it is clear that the earliest human occupation south of the ice happened long before the corridor existed, though it is not yet possible to date, or locate, the first instance, and these early camp sites are all dated to the two millennia before the 'opening'. (There is a recent claim of evidence for human presence in Mexico about 40,000 years ago – greeted with derisive scepticism, of course, by those wedded to the ice-free corridor theory.) This is not to say, of course, that later migrants did not use the corridor, when it was both open and capable of supporting them and the animals they hunted, though there is still the absence of archaeological sites along the corridor to account for. But it does mean that humans, settled in Beringia from about 14,000 years ago, fairly speedily found their way south, through or past the ice sheet; and the problem is not if they did it, but how they did it.

The Alaskans

The people we are concerned with here, however, are those who lived in Beringia, who were certainly blocked from moving south by the barrier of the ice until about 11,000 years ago. A group of sites in Alaska along the Nenana and Tanana Rivers has been investigated and shown to have been in use from about 11,600 years ago or perhaps earlier (Fig. 28). (There are other sites in the region which are even earlier than this, but it is these sites which are important here to show activity at the end of the Ice Age.) The region is in central Alaska; to the south the Alaska Range and the easternmost of the Aleutian Islands were at that time ice-covered – an extension of the Cordillera ice sheet – and this ice separated the central Alaskan area of these settlements from the Pacific Ocean, which here was at the time covered by pack ice extending out from the Cordillera ice sheet. To the north was another ice sheet covering the Brooks Range, and the great continental ice sheet was off to the east, not very far away.

These riverside camp sites were therefore in an area of tundra which was blocked off by high mountain ranges covered by thick sheets of ice in three directions. The only opening to other lands was to the west, to Beringia and Siberia, and this is presumably where the Nenana and Tanana hunters and their ancestors had come from. Life for the people was undoubtedly extremely hard, their numbers were few, and they had to keep moving in search of food.

Each of these Alaskan sites shows a sequence of relatively brief occupations spread over several hundred years. The food supply of the people came from the animals of the Arctic – bison, elk, moose, migratory birds such as geese and ptarmigan, small animals such as hare, fox, otter, and fish. Mammoths were apparently either not hunted or were not present in the region, but the people did use mammoth ivory and bones, which they scavenged from the skeletons of animals which had been dead for several hundred years. (Some of these may also have been still hunted elsewhere; but if one kills a mammoth one would consume it at the kill site, not lug it home over several miles of tundra, muskeg, and bog to a camp site which was only temporary anyway.) Only small indications of the habitations were found, indicating that tents were their shelters, or rock shelters where these were available to them; this, and the animal bones found, implies that the sites were occupied only during the summer. The occupants' winter homes have not been found, but were presumably in caves in the area, or possibly a long way away, even in area of the Bering Sea now flooded.

These people were hunters living the same lives as many of the other people living in Eurasia, all the way to Doggerland and Meiendorf and the French and Spanish caves. The land was tundra, cold, snow-covered in the winter for at least six months of the year, often permafrost at a depth of only a few inches below the ground surface, with a very scattered, thin population of human beings who camped briefly at chosen sites and soon moved on. The repeated intermittent use of the same sites over a long period implies that they always chose the same sorts of places in which to camp, and that some places were their favourites, and that they may have had a regular round of seasonal camp sites. It also suggests that the successive occupants were probably of the same extended family; that is, which is surely very likely, that the extended family, over several centuries,

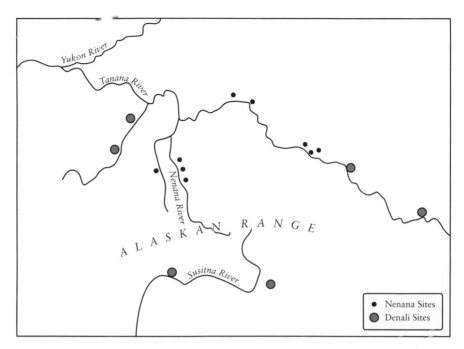

Fig. 28. NENANA AND DENALI SITES IN ALASKA. The archaeology of South America is largely divorced from that of the northern half of the continent, but discoveries there have steadily undercut the theory of the late colonisation of both. The decisive blow was the excavation at Monte Verde in Chile, which produced radiocarbon dates showing occupation a thousand years 'too early'. Since then several other sites have produced early dates, or, in other cases, disputed early dates which have become accepted. (There is even a possible presence of men as early as 30,000 to 40,000 years ago, but nothing is yet proved.)

circulated regularly through a familiar sequence of habitation sites with the seasons – so long as the living and hunting conditions remained much the same. Like their contemporaries in the rest of Eurasia they lived by hunting, and their prey provided them with meat, fur, bone for food and fuel and shelter, clothing, and with materials for tools and weapons – the essentials of their lives. In the winter they in effect hibernated, living on food gathered and stored from the summer hunting.

Archaeologists characterise these 'cultures' by their physical remains, especially taking account of the perceived differences between their sets of stone tools. Taking a more drawn-back view of the situation, however, the various groups do not seem all that distinctive. The Nenana complex is not really very different from similar groups in Alaska, which are dated a little later, but within the same general period. The campsites of these other groups were located in only slightly different places. So this 'Nenana Culture', which is dated between 11,000 and 10,000 years ago, and the 'Denali Culture', which overlaps it both geographically and chronologically (10,500 to 8,000 years ago) consisted of hunters who were living essentially the same life and using virtually the same tools. Both cultures were succeeded by the 'Northern Paleoarctic Tradition', and this shows very much the same characteristics, and in much the same geographical spread. These then are all remains of groups of a succession of people pursuing a similar hunting life, and probably all related to one another by descent. They therefore lived in small groups in the same area between about 11,000 and 8,000 years ago, a period overlapping the Last Glacial Maximum and the decisive retreat of the great ice sheets. These were the people whose descendants and perhaps whose sons, made the journey south into the 'Lower Forty-eight', once it became possible, but who will have discovered that they were not the first to make that journey. The question, again, is how they got there.

Moving South

From 11,000 years ago, while the people of those Nenana, Denali, and Northern Paleoarctic cultures were migrating and hunting around inland Alaska and Beringia, those low lands between Alaska and Siberia were being reduced in size by the rising of the seas, and the available land was also reducing due to the advance of the tree cover. The pack ice which had rimmed the old coast south of the Aleutian Islands broke up and melted; by 10,000 years ago the whole lowland between the two (modern) continents was under water and the land connection between them had been severed (the Bering Strait). Meanwhile the ice sheet covering the Rocky Mountains, the Alaskan Range, and the Aleutian Islands retreated, and the 'ice-free corridor' opened up, though until it was explored the hunters did not know that it might lead anywhere. For them it was, along with other areas newly free of ice, simply an area where they could now hunt, an area which grew in size rather more slowly than the land being lost to the sea was submerged. The effects on the inhabitants of all these changes were no doubt severe. It seems a reasonable conjecture that the people of inland Alaska had hunted over, possibly even lived in, the lost lands which were now flooded, and that their lives were now made much more difficult as a result of the disappearance of those lands, the restriction of the new trees, and the continuing ice sheets, even as more lands slowly appeared out of the ice elsewhere. The inhabitants had to choose their futures – some went north with the ice, and became Inuit; some south to become fishermen and whale-hunters, the Aleut (Chapter 2); others moved southeast past the ice into the great double-continent.

It is just at this time, as Beringia was vanishing, that the earliest sites of humans are found in what had widened from an 'ice-free corridor' into a much more accommodating and accessible alternative to the lost lands. The gradual and intermittent withdrawal of the Laurentide ice sheet from about 12,000 years ago was complemented by the rather more rapid reduction of the Cordilleran ice sheet, so the two ice sheets withdrew from one another, probably not for the first time in the erratic Ice Age. The ice-free corridor therefore widened, though the process took a good millennium to produce a viable route for the passage of families of humans. This change was accompanied by the relatively rapid reduction of the ice covering the Alaskan Peninsula and Aleutian Islands, and by the simultaneous disappearance of the pack ice. These developments not only facilitated the movement of people south into Canada by land, but also opened up a way from inland Beringia to the Pacific Alaskan coast. The people of the Paleoarctic Tradition were now able to occupy sites in both the Brooks Range to the north of the original territory (the Gallagher Flint Station site is dated to about 10,500 years ago), and sites along the Seward Peninsula (the Kvichak Bay and Ugachic Narrows sites) which were occupied by about 9,000 years ago. This, in fact, indicates that an alternative route southward had opened up, one which was much easier than the difficult 'corridor' (Fig. 31).

Two sites might show evidence of the movement into Canada by land. Charlie Lake in north-east British Columbia, east of the Rocky Mountains, is a cave site with just a little evidence of human activity dated to about 10,500 years ago; Vermilion Lake, further south in Alberta, is of roughly the same date or even a little earlier. These are dates which are a good millennium or more later than the 'opening' of the 'corridor' in which they are situated. These indications are of a fairly brief occupation,

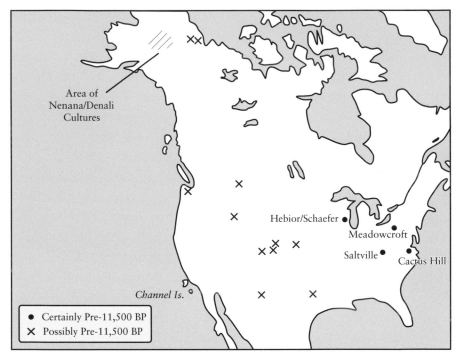

Fig. 29. EARLY SITES IN NORTH AMERICA. The controversy over the arrival of man in America is now resolving into an acceptance of earlier rather than later arrivals. At least five North American sites have produced dates before the end of the Ice Age; several others have been claimed to be early but the claims were dismissed; no doubt some of these will now be re-examined and accepted.

but they are well dated. They both might be considered to be remains of camp sites left by people moving north from the ameliorating land to the south as much as similar remains left by a small hunting band which was moving south as rapidly as possible, conscious of the ice to either side, camping briefly for a day or two, or perhaps a winter, before moving on, hoping for something better further on – or perhaps knowing that there were hunting grounds ahead because a scout had gone ahead and returned to bring the news. If the corridor really was being used, we must always assume that these movements were planned, and were not spontaneous or accidental – a family does not 'wander' into a corridor between ice sheets and move through a journey of 3000 kilometres by accident. Whole families do not explore; they send out explorers first. These dates are also roughly contemporary with the expansion of the

Fig. 30. EARLY SITES IN SOUTH AMERICA. These sites are no more than temporary camping places, probably re-used repeatedly, and possibly by the same family, for centuries. The Nenana sites are generally dated somewhat earlier than the Denali, but the two also overlap.

Northern Paleoarctic Tradition peoples into the nearby ice-free areas in both coastal and inland Alaska mentioned above. Yet two sites is not much. They are scarcely proof of the use of the corridor so early.

Whenever or wherever these people lived they had a hard life, short and difficult, just like all Ice Age hunters. Like the inhabitants of the dried-out North Sea at the same time, some of the people here developed the ability to build boats and fish, as the Aleuts demonstrated. With the theory of the ice-free corridor substantially deflated (though not wholly removed – it certainly existed, and it was almost certainly used, but it was

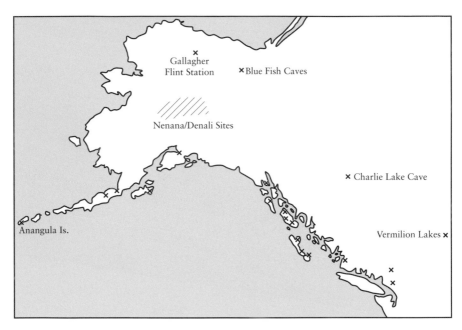

Fig. 31. EARLY SITES IN ALASKA AND BRITISH COLUMBIA. The relaxation of the grip of the ice permitted the expansion of settlements to and along the coast by descendants of the Nenana/Denali people. The string of sites along the coast requires us to accept the ability of the people at the end of the Ice Age to sail, at least in canoes, in the open ocean. The two northern sites on the map are a sign that the ice on the Brooks Range had gone; the two to the east show that it was possible, eventually, to travel south and that the 'corridor 'was 'ice-free'.

A considerable number of species of animals died out in North America at the end of the Ice Age. Five of them are pictured here. However, they went extinct at different times – the Short-faced Bear by about 12,600 years ago, the Giant Sloth three millennia later. There is thus no evidence of man being more than marginally involved, nor can it be really described as a 'great extinction'.

not an easy route, nor the exclusive one, nor even possibly the earliest one, towards the south), another theory has been produced to explain the colonisation movement from Alaska into the rest of America.

The Water Route

This alternative, or additional, theory postulates that it would be possible for canoeists to make their way along the coast of what is now the Alaskan Panhandle and into the fjords of the coast of British Columbia and its offshore islands, thereby bypassing the reducing ice sheet in the Rocky Mountains. The conditions must have been very similar to those in Norway at about the same time, with the ice still frowning from the mountains, icebergs in the sea, and with only small landing places available, though the numerous and larger islands will have helped. There would be few animals to hunt, much bare land without trees, and the necessary reliance on fishing for food. There is in all this, of course, an unspoken assumption, at least among American archaeologists, that this is what the people should have been doing all along, that they, as good Americans, were all obviously searching for such a route into what became the United States, and that the ice was foiling those living in Alaska of the chance to move south.

A series of archaeological sites has been found along the north-west coast, characterized in many cases by the use of 'microblades', the manufacture of much smaller and more delicately shaped stone tools and weapons ('microliths' in Europe; this is a matter I shall return to in the next chapters.) Two sites in the north produced these items; Beluga Point on the mainland coast, and Craig Point in the Kodiak Islands. These are dated fairly late, about 8,000 years ago, and by this time other sites along that coast had been occupied up to 3,000 years before; further south along the Alaskan Panhandle and in the islands off the coast of British Columbia there is a whole series of sites which have produced occupational evidence dated to 1,000 years earlier, about 9,000 years ago.

The sites in the islands and on the coasts, of course, require that the people who had inhabited them had the use of boats, since travel overland was very difficult. Some of the sites show that the inhabitants had the use of obsidian, which had to have been transported across the Straits from sources on the islands. The sites from Kodiak Island to Queen Charlotte Island are more or less contemporary, in the sense that they are in the same tradition, the 'Northwest Microblade', a successor of the Northern Paleoarctic, and the range of dates indicates a progressive occupation of this coast.

They show that the people lived in part – in large part, in some cases – by fishing, both in the sheltered waters and in the ocean. They had clearly reached their new homes by pushing through the Alaskan Mountains and moving south by boat along the coast. Whether they did this before others travelled along the ice-free corridor is not relevant here (though they all probably explored any newly available land simultaneously). Nor is it relevant that, as sites from California to Pennsylvania to Chile show, they were not the first to move from Alaska southwards. Here the issue is that both groups, by sea and by land, were reacting to the changes brought about by the end of the Ice Age, and the extent of the ice. The two groups were therefore first of all exploiting any newly available hunting grounds near their original homes with a view to increasing their resources.

Californians

Relevant here is the evidence from two sites in California. At Mostin, not far north of San Francisco bay, human remains have been found in a fairly large cemetery, some remains from which have been dated to over 11,000 years ago. As is always the case with very early dates in the Americas, everything about the find and its site, its dating and the excavation has been criticised in fairly savage terms, largely to ensure that the old migrations model which insists on no inhabitants before the end of the Ice Age remains intact. Its dating is certainly uncertain, even ambiguous. Further south, however, at Santa Rosa Island, one of the Channel Islands facing Santa Barbara, a site has also produced human remains and these have been carefully (and repeatedly) dated to about 11,000 years ago. This latter find is particularly convincing (and clearly lessens the effect of the scepticism about Mostin). It also means that the people in the area were using boats at that time, for these islands have never been connected to the mainland. The early date, though it is within the period of the putative beginning of the movement south, makes it much more likely that 'Arlington man' (the name given to the remains from Santa Rosa) was descended from the earlier inhabitants of the continent; their sailing skills cannot therefore be used to support the theory of movement by boat along the coast from Alaska. But it may be that this is a possible explanation of the means by which the earlier inhabitants pushed past the ice; the scenario would then be an initial coastwise arrival, whose repetition was blocked by an increase of the ice in the north in the Last Glacial Maximum.

Microliths

Technologically, the use of 'microblades' is an important development. It was more economical of materials, and the finished products could be used for a variety of tools, such as spear points, arrow heads, axe blades, and so on, sometimes interchangeably. The small blades could be more easily carried, and supplies of prepared replacements could be taken on the hunt, say in a leather pouch, rather than new stones having to be either carried or sought out when larger tools became blunted and useless through use. It was clearly a development out of the earlier cultures, but the crucial point for this chapter is that these sites are mainly found along the coast.

Conservatism in the Northwest Coast

The area in question, the Alaskan Panhandle and the coast of British Columbia, is a difficult region, not easily settled. As in Norway the mountains plunge to the sea, with no concessions to gentleness, but they do provide numerous islands and inlets and fjords where there are small landing places. This was a coast where, if anyone lived there, they were compelled to use boats; there were no other ways of getting about, or of getting there in the first place. Norway had been occupied some thousand years before in much the same way. The difference between the peopling of these two lands was that the microblade users from Alaska who moved along the north-west coast moved south out of Beringia, whereas the new Norwegians had moved out of flooding Doggerland going to the north. The north-westerners moved into better lands for their purposes; the Norwegians went deliberately into the same sort of lands that they had left.

The fishermen and boatman who settled the north-west coast of North America were moving away from the frozen lands of Alaska/Beringia and going to live in a warmer climate. Since others in Alaska stayed where they were (to become Aleut and Inuit) this was quite obviously a deliberate decision by the migrants: they were searching for a better life, with more abundant resources, and in a warmer land. Their new homes were probably richer in resources of fish and game and wood at least. By specialising in fishing they were able to colonise the coastal areas even while the Cordillera ice still blocked the nearby mountains.

These fishing societies were therefore intentionally acting as conservatively as had the new Norwegians. Their purpose was to find a land where they could continue living in the same way as they had in Beringia, but in better conditions. Like the Norwegians they found that they had to move to a new land, and had to specialize in hunting fish rather than land animals, though they never gave up hunting by land. They will have learned quickly enough that the skins of the largest sea mammals (seals, sea lions, whales) were admirably suited for making skin boats and even for their clothes, and their bones were as useful as the bones of land animals, while the forests of their new homes, when they developed them, were exploitable for the construction of homes and boats and for fires and making tools, as well as sheltering huntable animals. It was not so much a new life as one whose focus shifted to different resources in a different and more productive location.

By contrast, their cousins who moved north to become the Inuit, reached lands which were just as hard and cold and poor in resources as their original homes, while their other cousins, the Aleut, abandoned land hunting altogether. Still other groups continued on further to the south into further new lands. They may well have found that the islands and coasts were fully occupied, and so were effectively excluded by the earlier arrivals. There were, of course, only small areas which could be occupied in this land of sea and forest and ice covered mountains, and islands. At the same time the corridor, as the early sites at Vermilion Lake and Charlie Lake Cave may show, was allowing other human migrants in similarly small numbers to reach the interior of the continent. Until about 8,000 years ago they will have found the land near the ice subject to the flooding of the great glacial lakes, but going on further south again they reached the Great Plains, an area subject to the dust clouds driven by the dry winds. These were populated by the large native animals, however, mainly herbivores of a similar sort that they had hunted in the north, and this would certainly be sufficient to provide them with food. They also found that the land was already populated, if thinly, by earlier human arrivals (see Gallery IV).

Spreading into the New Land

Not more than a few centuries after the beginning of the movement along the coast and along the corridor, a large number of occupied sites are known in the area of the continental United States. These are allocated to several 'cultures' by the archaeologists, some with technological names – 'Western fluted point', 'Western stemmed point' – others named after findspots – 'Clovis', 'Folsom' – but which all imply the same sort of hunter-gathering lifestyle, and broadly similar to the Nenana and Tanana of Alaska. Their prey here were any animals which were available, at times mammoths, but also mastodons, bison, pronghorns, and other large beasts, and the life was, as ever, migratory.

From now on it is possible to distinguish a steadily increasing number of these 'traditions' and 'cultures' and 'complexes', as the people spread through the North American mainland, and on into Mexico, and as the numbers of people increased, and as their hunting grounds became defined and limited. They met and no doubt intermarried with (or fought and killed) the small number of hunters already present. The social stage was therefore set for the development of the tribal lives of Amerindians throughout the later North America.

The two streams of people moving south from Beringia were separated geographically in their migrations by the Rocky Mountains, which remained ice-covered as far south as the borders of British Columbia and Washington State for a long time, and there were further extensions of ice along the southern Rockies, which were barely penetrable for long distances. Not surprisingly, given their different roots, routes, and origins, the two streams led on to rather different societies. In Oregon and California the hunting bands were always small, but on the coast the fishermen flourished in such benign places as the San Francisco Bay area, the Puget Sound archipelago, and the Channel Islands off the coast of southern California. The new inhabitants used the plentiful resources of the sea in all these areas – shellfish and fish, notably sturgeon in the Bay area, and a wider variety, including swordfish, in the south. In the Oregon area at least one society developed great skill in whaling, and others specialised in gathering salmon in the vigorous rivers.

Extinctions

In the great interior the new people arrived at a time when great changes were going on as a result of the change in the climate, and these changes continued for several millennia. The shrinkage of the ice sheets and the draining of the Ice Age lakes were the preliminaries to even greater changes, especially in the vegetation, as was the case also in Europe. The oscillations of the climate, in part as a result of the great pulses of water draining from the interior, was another factor. The rising temperature thawed the permafrost and helped the land to dry out, and at the same time it encouraged the spread of forest, first of all pine trees, and later mixed deciduous forest in the east, and in the west the great Douglas firs. In the centre of the continent the climate remained wetter than today for some time, and forest spread while the land was still wet, but it became drier with the removal of the ice sheets in the north. The trees then became confined to the river valleys, and the rest became the grass covered plains which existed until the new settlement of farmers from the south and east.

People arrived from the north, and so did animals. In the new land they found other animals with which they were unfamiliar (Gallery IV) – an echo of the experience of the first Australians in meeting such animals as the kangaroo 50,000 years before. In the 2000 years or so following the start of the retreat of the ice a number of these animals became extinct. Why this happened is not known, though theories blame both men and the environment for it. Both explanations are difficult to accept on their own. Men were scarcely numerous enough to kill off all the mammoths, mastodons, camels, several genera of deer, and other animals, and several carnivores as well, including a bison and a wolf, a sabre tooth and a huge bear, throughout the present area of the United States, in only a couple of thousand years. And if men did the work, they were unusually thorough, for all these animals completely vanished, leaving none in isolated refuges or uninhabited regions, as had happened elsewhere and in the past. (Note the history of the Pacific sea otter, driven almost to extinction, except for a refuge at Big Sur in California, whence the animal's recent revival.) Are we supposed to believe that men deliberately sought out every last one of these animals in order to liquidate them all? Such a theory is untenable because it does not take account of human behaviour, quite apart from the impossibility of the task.

On the other hand, the changing environment can scarcely be blamed either, for this change took place slowly enough for the animals to adapt to it and to move on to the new grazing lands – just as the mammoths did in Eurasia – though the increase in the temperature may have had a particularly damaging effect. Nor is an acceptable combination of the theories convincing. The only fact available is that the 'great extinction' happened – though it is no 'greater' than others of its type; one has the feeling that American scientists are proud to have had another 'great' event happen in their land. It was, in fact, one of several such events which have occurred at different times and in different places in the past (and the present), and this implies that, like global warming, it was largely a natural phenomenon, one to which the Stone Age hunters contributed only a little. America was no more exceptional in this than in any other matter.

The New World

The hunters found themselves in a vast double-continent, which was warmer and infinitely more varied than the cold desert tundra and ice in Beringia which they had been used to. The point, of course, is that we see here a different version of the escape from the trap of increasingly warm temperatures as was seen in Europe. The rise in the sea levels of the oceans had the same effect in Beringia as in the North Sea, forcing the people to exist in more limited hunting territories, and concentrate on fewer animals in their hunts, as the low-lying land they had used earlier was gradually flooded, forcing them also to move or adapt. The direct land connection between Alaska and Siberia was finally severed about 10,000 years ago, by which time the ice-free corridor was just about open, if difficult, and the microblade-making people of the North West had taken to the sea and the islands and had spread south as far as Oregon or even California. The great ice-dammed lakes drained away, the great lake of the Great Basin slowly evaporated, shrinking to the Great Salt Lake, the ice melted. The big-game hunters of the Clovis culture spread over the Great Plains and the western mountain regions; others were able to move on to the less productive lands of the north once they were free of water and ice; still others continued along the coast of the Far West, and some moved into the lands of the future deserts of the Southwest. These were hunters of a slightly different technological tradition from those of the Plains, but still very similar in their culture. Still others stayed in place in the north, to become the ancestors of the Aleut and the Inuit.

The reaction of the Sibero-Alaskans, therefore, to the global warming over the millennia from about 11,000 years ago, was to move southwards, out of their newly constricted and decreasingly productive homeland, and into a new land where they found that their lives had to become rather different than they had been in Beringia. This was not their intention, of course, but the conditions in their new lands enforced change on them. More and bigger animals were available to be hunted on the Plains, but the change, like that for the new inhabitants of Norway, was one of hunting grounds only, not of their lifestyle; the new inhabitants of the North West similarly moved their hunting grounds into the sea and hunted fish and marine mammals, just as they had in the north. But it was not the same, for the land was drier, or wetter, or hotter, or bigger, or the animals were different, and life had to change to meet these new conditions; they had, that is, to adapt, willy-nilly, to the new life.

The object of all these people had been to continue to live in the same way as they and their ancestors had done for millennia. This makes sense, since this was their experience, and any experimentation was quite likely to bring a disaster. To stay the same, however they found they had to move, from Doggerland, as it disappeared, into Norway, from Central and Eastern Europe into Siberia, from Beringia into Canada and southwards. And in their endeavours they were generally successful. Yet movement from one land to another really is a drastic change. Some groups, on the North-West American coast and in Norway, found that they had to concentrate their work in a different direction, fishing rather than hunting land animals, for instance. And the big-game hunters in both Siberia and North America found that, largely by reason of changes wholly outside their control, their primary prey eventually died out.

All this was forced on the people involved by the relatively rapid changes in the climate at the end of the Ice Age. Within about 3,000 years the Scandinavian ice sheet had gone, the Cordillera ice sheet had melted and the Laurentide sheet had been reduced by half, and then it suddenly collapsed, Lakes Bonneville and Missoula vanished, Lake Agassiz/ Ojibway drained away into the ocean. The temperature in Europe and the southern half of North America rose by an average of 7°C or 8°C. In terms of the life of any single human there was little enough to notice in the changes, for it took longer than a human generation to become noticeable in most cases – except in the catastrophes; in terms of the collective memory of the clan there was pressure on the people to adapt to the new conditions, to aim to retain their old lives. Wherever they migrated to, if the conditions they found were not the same as they were used to, they adapted by changing their lives to make the best of it, or to survive, though being human, they changed as little as possible. And that 'best' was usually an improvement on what had gone before – better climate, more animals to hunt, warmer. By moving to expand their hunting grounds, in compensation for losing their Beringian lands to the sea, the Beringians were in fact escaping from the grip of the ice, though that was probably not their overt or original intention. So one of the effects of the changes in climate at the end of the Ice Age was now to force men to become used to change, to become even more adaptable, and to look outside their traditional existence for ways to survive and prosper.

Further Reading

Changing environmental conditions in North America are described by Pielou, *After the Ice Age*, cited in Chapter 2. The discussion of the 'peopling' of North America can be followed in Thomas C. Dillehay, *The Settlement of the Americas*, New York 2000 (Dillehay was the excavator of Monte Verde, and he looks most attentively at the evidence from South America) and in E. James Dixon, *Bones, Boats and Bison*, Albuquerque 2001, who approaches the issue from the perspective of his excavations in the north. An entertaining account of the Meadowcroft excavation, and the tumult it and the Monte Verde evidence caused in the archaeological dovecotes is in J.M. Adovasio and J. Page, *The First Americans, In Pursuit of Archaeology's Greatest Mystery*, New York 2002 (a title all too typical of the self-centredness of United States' researchers); on the controversy see also David J. Meltzer, *First Peoples in the New World, Colonising Ice Age America*, Berkeley and Los Angeles 2009. For the history of Lake Agassiz, see D.W. Leverington *et al.*, 'Changes in the Bathymetric Volume of Glacial Lake Agassiz between 9,200 and 7700 C14 yr B.P.', *Quaternary Research* 57, 2002, 244–252.

Much work has also been done on the sites of people who came to live along the Pacific coast of North America. See in particular R.G. Matson and G. Coupland, *The Prehistory of the North-west Coast*, San Diego CA 1995, W.R. Hildebrandt and T.L. Jones, 'Evolution of Marine Mammal Hunting: a View from the Californian and Oregon Coasts', in the *Journal of Anthropology and Archaeology* 11, 1992, 360–401, J.M. Erlandson and T.C. Rick, 'A 9,700-year-old Shell Midden on San Miguel Island, California', *Antiquity* 76, 2002, 315–316, and T.C. Rick et al., 'Paleocoastal Marine Fishing on the Pacific Coast of the Americas: Perspectives from Daisy Cave, California', *American Anthropologist* 66, 2001, 595–613.

Gallery IV

Extinct Animals of North America

Periods in which animals have become extinct are not unusual. One took place in Australia about the same time as that in America – but being in America, it has to be a 'great' extinction. One may also note that it happened during and at the end of the Ice Age to several varieties of humans, including the Neanderthals.

The extinction in America did not happen all at once, but was spread over at least 3,000 years, as the examples shown here indicate. These are not the only animals to disappear, but are perhaps the most spectacular. The reason is not fully understood, but the change of climate at the end of the Ice Age and the increased warmth were no doubt part of the cause.

IV.1 Woolly mammoth, extinct by c.10,500 years ago.

IV.2 Mastodon, extinct by c.10,400 years ago.

IV.3 American lion, extinct by c.10,400 years ago.

IV.4 Giant short-faced bear, extinct by c.12,600 years ago. (*Dantheman9758 via Wikimedia Commons*)

IV.5 Pronghorn deer, extinct by c.11,300 years ago.

IV.6 Dire wolf, extinct by c.10,000 years ago.

Chapter 4

Floods and Fish

The rise in the general sea level was as important a result of the increase in the global temperature as the shrinkage and eventual disappearance of the great ice sheets. Like the melting, however, the water level rise was not uniform in its effects, nor in its timing. The disappearance of the great ice sheets permitted the land they had covered to be uplifted, the process of isostatic recovery which has been already referred to; the erratic changes effected in western Scotland by the combination of these two have been mentioned already. In Scandinavia, the land is still rising in relation to the sea to this day, as it is in North America. Events like the sudden emptying of the great ice dammed lakes in North America had various effects on the climate as a whole, both immediate and in the longer term, and other events like the flooding of low-lying lands in several regions, by siphoning off large amounts of water, would probably retard the rise elsewhere. All these influences meant that the effects of the changes in sea level inevitably varied from place to place.

Flooding

Several specific regions were subject to particularly extensive flooding by the rising waters, in addition to all the coastlands which were inundated. The contrast is exemplified by what happened in Australia. There the steep eastern and western coasts were relatively little affected, though the coastline usually 'receded' twenty or so kilometres at a minimum. In the north, on the other hand, the lowlands separating Australia and New Guinea were flooded, as was the Gulf of Carpenteria, so that in the north and north-west a large area of low land became sea. Tasmania became separated from the continent by the new Bass Strait. As a result, the original single great continent became divided into three large islands and numerous smaller ones. Perhaps fifteen per cent of the land disappeared, 2.5 million square kilometres. Such a variation clearly had

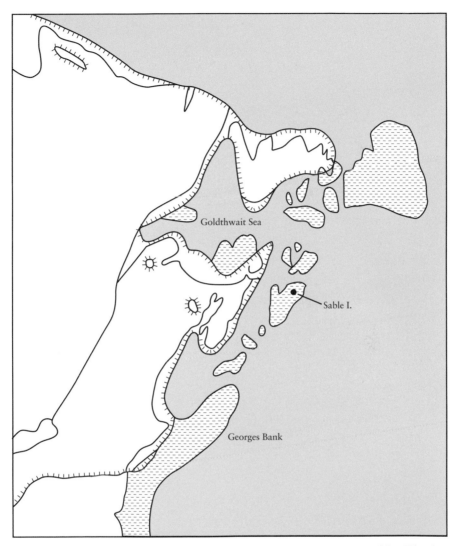

Fig. 32. EASTERN NORTH AMERICA UNDER THE ICE. The ice blanketed the present land, and the lowering of the sea level brought new lands to the surface. This is the sort of change could be seen wherever there is a shallow continental shelf. It is obvious that if there had been human presence in this area in the Ice Age the evidence would have vanished, either to the sea but under the sea.

a disproportionate effect on the people of the different regions. (Other drastic changes took place in Australia, to be discussed in Chapter 6.)

The North Sea and the English Channel was another of these flooded areas, of course, and the results there have been noted already in the

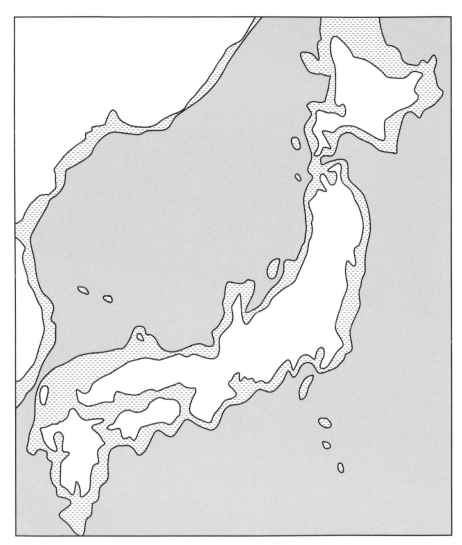

Fig. 33. JAPAN BECOMES AN ISLAND. As the sea level rose Japan's connection with the continent was severed, first by a narrow channel, then by a much wider strait. But the sea continued to rise, and flooded the lowest lands, so that the single island broke up into an archipelago.

discussion of the peopling of Norway (Chapter 2). Beringia, or rather its disappearance, has also been discussed, with substantial results for the people there, and for America as a whole (Chapters 2 and 3). Elsewhere in North America the rise helped to create the complex of peninsulas and archipelagos along the north-west coast which became the home for the

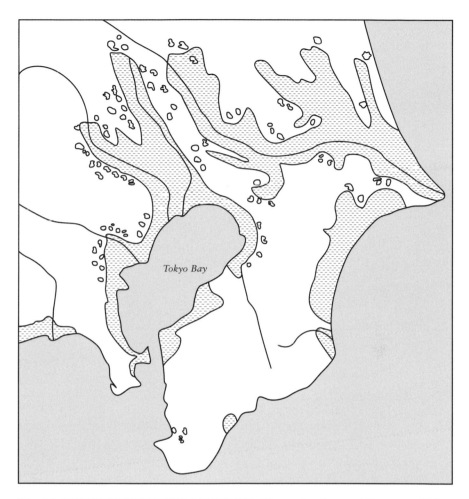

Fig. 34. MESOLITHIC SHELL MOUNDS. The sea level around Japan rose higher than at present, then fell back. Considerable areas of the present Japan were thus flooded for a time. In the Kanto, the area around Tokyo, this phenomenon has been mapped by locating the positions of shell mounds built up at the time of the high sea level. But the quantity and location of the shell mounds – the remains of foragers collecting shellfish – also indicates the density of the occupation of the area, and the static nature of Mesolithic Japanese society.

early population who moved – or were driven – south out of Beringia/ Alaska. On the east coast of North America large areas of lowland became flooded – the Grand Banks of Newfoundland in particular ceased to be land and became a shallow sea – and the coastlines of the Gulf of St Lawrence, Nova Scotia, and Maine were changed extensively as the ice retreated (Fig. 32). Japan became isolated from the continent by the flooding of the land bridges which had connected what had been a peninsula in the Ice Age to the continent, and as the waters rose still higher, the original large single island became broken up into the present archipelago (figs. 33 and 34).

The greatest single change in geography, however, took place in Southeast Asia. The long fat peninsula, which in the Ice Age had stretched south and east from the present continental mainland as far as Sulawesi, and which the paleogeographers have called 'Sundaland', was composed in large part of low-lying lands joining mountainous regions. These lowlands were now flooded to form the seas and straits separating the modern Indonesian and Malaysian islands and peninsulas. Probably half the land between Thailand and New Guinea disappeared, and the seas between these lands became well scattered with large and small islands (Fig. 3). This had happened by 11,000–10,000 years ago, by which time the ocean level was a little lower than it is now; some further, but slower, flooding continued for a time.

Loss of Evidence

All this change has probably hidden a good deal of archaeological evidence of life before the flooding. Any people who lived in the now-drowned areas were themselves drowned, or, like the people of Doggerland, were driven out into the surrounding lands – though some will also have died. The archaeological remains of their lives are now hidden under the waters. It is only recently that any serious underwater archaeology has become possible, and then only in shallow areas and close to coasts, so little or nothing has yet been discovered – the search for Doggerland took years. Ultimately no doubt it will become possible to search the wider shallow seas, and maybe even find ancient campsites, though much of this evidence has probably been buried deep in the mud as well as being under water. Meanwhile the flooding of the shallow seas provides a perfect excuse for not knowing what happened and who lived in those areas – 'the evidence has been lost when the lands are flooded' is a perennial complaint/excuse – and thus there is left a substantial loophole for speculation.

(It is, however, necessary to be wary of this easy temptation, for to suggest that since the evidence has 'almost certainly' disappeared under the sea, then it is legitimate to speculate about what that 'evidence' is, and what it might mean. As it became clear that Doggerland existed for some millennia and was inhabited, it has become the repository of notions about the spread of ideas and people between the continent and Britain, or even as the place where those ideas were originated. It has already rather too frequently appeared in these pages in that role – in the speculation about the colonisation of Norway I have indulged in, for example. A similar mind-set appears in coastal South America and to some extent in North America as well, where it is sometimes assumed that the evidence for the maritime migration along the Pacific coast has simply been drowned. It has also been used as an explanation for the absence of evidence for human settlement in Scotland – but, as is often said, absence of evidence is not evidence of absence; all such comments are speculation or mere guesswork, or the exclamations of frustrated archaeologists.)

The Gulf

One other large area which went from being land to going under water was the Arabian/Persian Gulf. This is now a shallow sea, in effect an inlet of the Indian Ocean, much of it less than thirty-five metres deep, and so it was largely dry at the height of the Ice Age when the sea level was lower than now by up to 115 metres or so. An elongated Tigris-Euphrates River flowed along it, and was joined by other rivers flowing from the Iranian mountains, the whole emptying into the Indian Ocean at the Gulf of Oman. Details are scarce, but it is probable that the area was flooded from the ocean at some time before 9,000 years ago – after the settlement of Norway, but about the time the Beringians were moving south, and before the great '8.2 KYA event' threw great quantities of water into the world ocean. Its shallowness means that the flood would have been fairly rapid when it happened. There may have been humans living in the area before the flood, though it seems generally agreed that the area was subject to an arid climate until at least the time of the flood, with the rivers the main sources of fresh water, and so the river valleys were no doubt lined with trees (Fig. 35). Since it was flooded those rivers have been depositing silt

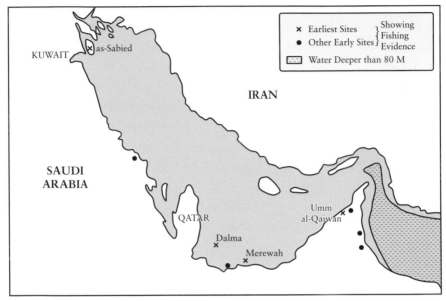

Fig. 35. EARLY FISHING IN THE GULF. Most of the Gulf is shallow water, with only small areas deeper than 80 metres. The rise in sea level by 100 metres at the end of the Ice Age was therefore crucial to its formation. There was no Gulf before the great warming, though there may have been a lake. The relatively flat sea bed implies that the flood spread quickly once it began. The sites marked are those which show evidence of early reliance on fishing as a livelihood.

in the Gulf, and so gradually reclaiming it for the land. (This sea is an obvious candidate for underwater archaeology, if anything could be done in those busy and polluted – and political – waters; some of the results of that work are the origin of this section.)

The flooding of the Gulf would, of course, help to increase the humidity of the surrounding lands, a change which certainly happened about this time, though this was mainly the result of a temporary climatic shift, which did not last. From about 6,000 years ago the humidity faded away once more, so that by about 5,000 years ago the climate has been described as 'hyper-arid' – that is, the climate had settled down to its present condition of a general absence of rain, so that the land on all sides of the Gulf is desert, with rainfall arriving only occasionally, intermittently, and unreliably. In all likelihood, in fact, the lands have never really ceased to be desert, for even during the 'humid' period, the humidity then was much less than drenching.

The essential change which took place, therefore, was the flooding of a wide expanse of low land, through which the rivers had long flowed. A couple of areas are deeper than the rest, and may well have formed lakes before the flooding, or perhaps more likely dry troughs, but the level of any water in these areas was higher than the sea level during the Ice Age, and so these areas may have been merely intermittent lakes. Any people living in that land as the water flooded in from the ocean probably had plenty of time to get clear. They were, of course, hunters, and the rivers will have provided water and vegetation along their courses, though the general aridity will have drastically restricted the possibilities of success in their hunts. They will have moved out of the area as the animals moved away, but since the land was generally dry, their numbers (both of people and of animals) will always have been few. The flooding will have pushed them away from their old hunting grounds, but only as far as the shores of the new sea. These people were in all probability the ancestors of the people who are later discovered living along the Arabian and Iranian shores of the new inland sea which was the Gulf.

Examination of the remains left by the people who lived along the new shores shows that they camped in flimsy shelters and made their living in large part by fishing. The bones they left at their campsites are 80% or 90% fish; the rest of the bones are almost entirely of desert animals, such as goat and gazelle. The gazelle bones indicate that the people were hunters as well as fishers; the goats have been assumed, possibly wrongly, to have been herded, though they were also no doubt the hunters' prey at first; domestication came considerably later. This fish diet cannot have been what the pre-flood population lived on; in fact, it is likely it was exactly the opposite, mainly meat, and perhaps no more than 10% fish, caught in the river and perhaps the lakes.

Investigations have concentrated on sites located along the Arabian shores of the Gulf, from Kuwait to Oman; much less has been done on the Iranian side, though, as will be seen, the same conclusions can be reached as to the early inhabitation there as in Arabia by using a different set of evidence. There is, however, some useful evidence from islands in the Gulf. The earliest dated sites so far located are from Merewah Island in Abu Dhabi, and near Umm al-Qaiwan, the chief place of the small emirate of that name. These two earliest sites are of the 6th and 6/5th millennia BC, that is, about 8,000 years ago. The date of the flooding of the Gulf was perhaps 1,000 years earlier than that, and it had reached its present level by about 7,000 years ago, so these early sites were founded by people who were coping with the slowly rising sea level. From these earliest sites onwards until the recent past – oil has changed things in the past generation – one of the principal ways of life in the area has been fishing.

This was not a way of life which could have existed in the area before the flood waters came in, since there was no sea – though a much lengthened Tigris-Euphrates flowed through the arid land, and no doubt hunters who lived there, for there were some, caught fish when they could. (There may have been fishermen exploiting the Indian Ocean, but every area where they might have lived has been flooded – the rise in the sea level once again takes the blame for our ignorance.) When the sea rose, however, it did bring with it a new supply of food which had not been available to those living in the area before – fish from the sea. And the new sea turned out to be a rich source of fish, while its shallowness meant that it was relatively easy for the fish to be caught.

The remains even from the oldest sites show that sea fishing took place, and so effective fishing tackle and nets had been invented, and quite possibly boats as well. At Merewah Island the earliest hearth dates to about 8,000 years ago, and even in the scanty remains from that time, the bones of sawfish and sea bream were found, as well as remains of crabs and molluscs, which could be gathered along the shore. The other early site, near Umm al-Qaiwan, has more fish remains, but from a somewhat later context, though the site was still earlier than the Ubaid period villages in Mesopotamia, which began to develop about 7,000 years ago. (This is the archaeological period which marks the beginning of a major development of the wealth of the Babylonian area; one result was the extension into the Gulf of Ubaidian economic influence.) The remains came from a layer of burning underneath (and so earlier than) Ubaid-period materials, and included the same sorts of crabs and molluscs as at Merewah, but there was a wider variety of fish present – the remains of sawfish and sea bream were here accompanied by the bones of groupers, snappers, a fish like a sardine, barracuda, and tuna or mackerel, which are larger fish and probably could only be caught at some distance from the shore – hence, probably, from boats, marking both a greater ability at sea, but also a much greater expansion of fishing technology.

Many of the smaller fish could in fact be caught from the shore, and many of them were small individuals; this was no doubt in part an extension of the collection of the crabs and molluscs which feature prominently in the remains. Even fishing from the shore, however, required lines and nets and baskets, and for anything further offshore than a few metres, boats were certainly required. Indeed, the site at Merewah was actually on an island, which requires that the inhabitants needed boats for access to the mainland, which is now ten kilometres away, though it was perhaps less then. Dalma is another island off the Abu Dhabi coast, forty kilometres from the mainland, and on that island there is a site which has produced the remains of even more varieties of fish than at Merewah, and which was clearly the base for fishermen who sailed the length of the Gulf. It also produced pottery of the Ubaid type, exported from the early urban cultures which were then developing in Iraq. By then, the fifth millennium BC (7,000 to 6,000 years ago), it is clear that the Gulf was already a well-used waterway, with trade by sea developing. One of the ingredients of that trade was the boats of the fishermen of the Arabian shore who were well placed to act as intermediaries between Oman, the source of copper, and Iraq, the great and wealthy market for all sorts of raw materials, whose product at first was grain, and later all sorts of manufactured goods.

This fishing and trading lifestyle was stable and satisfactory for its practitioners, and for the inhabitants of the shores of the Gulf. The archaeological research has been done mainly along the Arabian shores of the Gulf, where a similar set of remains has been found at many sites all the way from Kuwait (at as-Sabied) to Umm al-Qaiwan almost at the mouth of the Gulf. The life involved a heavy reliance on fish of various sorts, including the crabs and molluscs of the shoreline. The settlements of the fishermen were established close to the coast or on islands in the sea – the Gulf is littered with islands – where their flimsy huts were built as shelters; of these the remains are often little more than patterns of stones on the ground, or simply the remains of hearths; long inhabitation of these places has resulted in the constant re-use of stone and wood and bone; bone was used as fuel, as burnt remains testify, and wood was always extremely scarce, having usually to be brought in from elsewhere – trading again. Given these circumstances, and the likelihood of the removal, or re-use, or destruction, of earlier remains, it is astonishing that the archaeologists have been able to find any remains at all.

The diet of these people was varied by the addition of the meat of goats and sheep (whose skeletons are archaeologically indistinguishable at this period) and gazelles. They also ate what vegetables and fruit, especially dates, they could find and grow, and no doubt they consumed vegetation from the sea when it was edible and available, but, judging by the bone evidence, the marine content of their diet must have been perhaps eighty percent of the whole. Their trading activities meant that they also acquired goods, and probably other foods and supplies of wood, from elsewhere once their participation in the trading system in the Gulf grew. But the fishing lifestyle developed before the trade developed. The participation in trading only became possible as an addition to the lives of the fishermen with the growth of wealth elsewhere, particularly in Iraq. The people of the Arabian shore and the islands had little surplus to trade and produced few tradable items, though some products, such as pearls, would be found and exploited later. Their profits from trade were those of the middlemen – transporters, sailors, and ships' captains – not producers or manufacturers, nor even merchants.

The Gedrosian Fish-eaters

The previous summary is based on the Arabian remains, but it was a style of life which also developed on the Persian side of the Gulf, along the Iranian coast to the east towards India, and in parts of the Red Sea as well. The evidence from the Persian side is not archaeological, but descriptions by historians of the expedition of Alexander the Great, whose army passed along the coast in 325 BC, and it requires a certain reach of faith to apply the information back to the earlier millennia.

In a fit of hubris Alexander brought a large part of his army along the coast of Gedrosia, a desert region of Iran facing the Indian Ocean, assuming that his command and organisational powers would get everybody through without damage. But the resources of the area were sufficient only for the inhabitants, and were wholly inadequate to feed his tens of thousands of men and their dependants. The men starved and went crazy from thirst, or were killed in flash floods, and many of the dependants suffered even worse. They did note, however, the existence on that coast of the Ichthyophagoi – the 'Fish-eaters' – people who lived by fishing from the shore (Fig. 36).

Another part of the army, under the command of Krateros marched by a more inland route, through lands fully capable of feeding his detachment, with orders to rejoin the king further on, and a third part, under Nearchos the Cretan, sailed in ships specially built in India from the mouth of the Indus River as far as Babylonia, coasting along the way. The sailors – actually men from the army with some maritime experience such as Nearchos himself – became intimately familiar with the Fish-eaters, whom they encountered repeatedly when landing to replace their supplies, usually without success. Nearchos' account was published soon afterwards and was summarised five centuries later by the historian Arrian in his *Indike*, a description of India. The geographer Strabo also summarised the geographical elements of Nearchos' account in book 16 of his world geography.

Strabo's version is the more revealing of the two. He describes an arid land, where the only vegetation was palms and thorn bushes and tamarisks – the palms being cultivated and concentrated in the few villages. The shortage of water meant it was impossible to cultivate the usual food plants, with the exception of the date palms, groves of which were to be seen at most of the villages. One of Nearchos' stopping points was a town larger than usual where for the first time for many days he thought there would be bread, for he could see areas of cultivation in the town. It turned out that there was only a very little wheat or barley bread, though he went to the trouble of capturing the town to get access to it; he found that the inhabitants regarded wheat bread as no more than an occasional delicacy; their own 'bread' was a preparation of meal ground from the flesh of baked fish. So their diet, one way or another, was largely fish, prepared in various ways from raw to baked to ground into meal; they also ate dates, and the meat of sheep and goats; but the animals had been fed on fish, so that even the mutton tasted of fish.

This is not, of course, exactly the same set of foods as the original post-flood fishermen in the Gulf will have had available, though it cannot

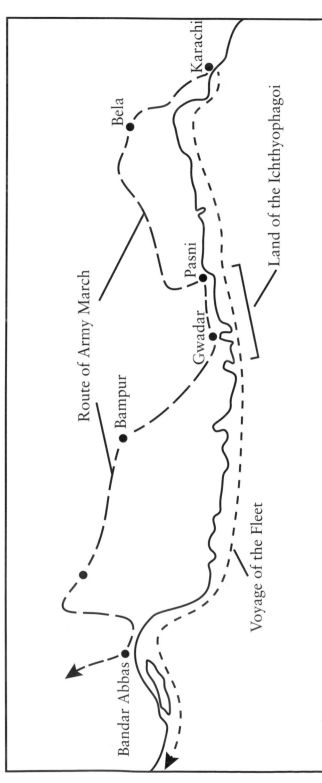

Fig. 36. ALEXANDER AND NEARCHOS IN THE DESERT: The intention of this dual expedition by land and sea was for the fleet to transport supplies, which the army on land would use for survival. But Alexander's march took him inland, out of touch with Nearchos and the ships. They met up at Pasni in the land of the Ichthyophagoi. This may also have been the 'city' which Nearchos attacked. Alexander, needless to say, seems to have blamed Nearchos for his own failure.

be far off. Clearly there had been some developments which eased the heavy reliance on fish. The palm groves at many of the villages provided a good alternative source of food (and this would have been available in its wild form even at the beginning); like the people of the Arabian shore of the Gulf and its islands, the Ichthyophagoi could acquire exotic foods – bread! – from elsewhere by trade; no doubt by Alexander's time their meat came from their own herds of sheep and goats. But the difference in diet was a matter of degree only. Fish in its varied preparations remained their basic food. It was a reliable enough source so that the people lived in permanent settlements and had some surplus – as at the town which Nearchos took – but hardly enough to be able to trade it, and certainly not enough to supply an army of several thousands of men dropping in unexpectedly. The whole country in fact resembles in many ways that of the fishing Arabs of the Gulf, though the geographical situation of Gedrosia prevented the people there from taking part in the trade which had been the source of much of the wealth of the Arabs; and yet the Gedrosians had established a stable lifestyle without it.

The villages were, Strabo reports, built of whale bones, the ribs being used as beams and wall support, the jawbones forming the doorposts – very like, perhaps, the Palaeolithic huts made of mammoth bones which have been found in the Ukraine. Oyster shells were also used, though just how is not clear, perhaps fastened to the bones to make a more solid seeming wall. The vertebrae of whales were used as mortars in which to grind their baked-fish meal. There is in Arrian's *Indike* no indication that whales were actually hunted, for both Nearchos and Strabo are emphatic that the boats the fishermen used were very frail and small and barely seaworthy; Nearchos was an experienced sailor, and his opinion is clearly reliable in this. They also noted that the Gedrosians did not row these boats, but propelled them along by a single pole, wielded from side to side as in a canoe, or, as he notes, as in a river, presumably with the oar hooked into the stern. Such vessels could hardly be used in a whale hunt. The bones must have come from the skeletons of whales which had been stranded on the shore in the same way as Inuit fishermen and the Beringians foraged for the bones of long-dead mammoths. Arrian, summarising Nearchos, notes that the Fish-eaters did not even use their boats for fishing, but that they fished from the shore, casting nets made from twine which was made from the bark of the date palm, though this was a supplementary method. Their main means was to fish the pools left by the retreating tide, when they netted any fish left behind – and ate some of them at once, raw (and perhaps still living), even as they emptied the nets.

These descriptions fit very well with the archaeological researches on the Arabian shore sites. The place is different, the time is several thousand years later, but the lives of the Fish-eaters of Arabia and the Ichthyophagoi of Gedrosia do not seem to have changed much over the millennia. It seems clear that the people, on both coasts, 7,000 years apart, were actually living essentially the same life in the same way. The Arabians had more access to outside materials, and took advantage of being on a trade route: the Gedrosians of the Iranian coast often showed fright at the appearance of the Macedonian ships, which suggests that they were unfamiliar with seagoing craft – though the Macedonians themselves were probably pretty frightening, and no doubt word had spread along the coast of their depredations elsewhere.

By the fourth century BC, when Alexander passed along that coast, the Fish-eaters had acquired several items of equipment which had clearly been unavailable to the Arabians of the Gulf 6,000 years earlier. They had earthenware pots for storage, and some of them sowed small fields of catch crops (which Nearchos mistook as a possible source of food for his men); they had palm groves to supply dates, a most nutritious food, and this was a source of food which had been bred up from the wild plants to provide a substantial crop. But they had no iron, as Strabo specifically notes, and their weapons were sticks hardened in the fire – possibly just as useful for spearing fish, and more easily manufactured and maintained than metal weapons – if they could get the wood. Clearly these additions to their lives were not essential, and they had carefully selected what they would add to their repertoire of tools and weapons – iron, for example, was of little use in fishing, and would rust away quickly in the sea-salty environment of the Gedrosian coast. Their houses, equipment, hunting methods, and general lifestyle, were very like those which their remote ancestors had developed when the Persian Gulf was first flooded and so became available as a source of food.

This therefore was a style of life which provided adequate, if perhaps monotonous, food for the inhabitants, who were using the materials available to them in their land in the most efficient way, wasting little, and carefully avoiding reliance on any material resources which were not available locally. By the time Alexander and Nearchos came by, this had been the sort of life lived in the region for at least 6,000 or 7,000 years, and there was no obvious reason why it should have changed. It was a life which had evolved to meet the new conditions produced by the rise in the sea level and the flooding of the Gulf at the end of the Ice Age, and which exploited the rich resources available in the sea. The ancestors of the people who stared at Alexander's men and ran away from Nearchos' ships, were perhaps some of the earliest to establish a fully settled and sedentary way of life anywhere in the world.

They may be compared with the Norwegians and the North-west coast dwellers and the Aleuts in North America, who had also resorted to the sea, and found that they had to settle in one place in order to exploit efficiently their new and rich resource. Those who went inland turned to – or continued – hunting land animals, on the other hand, were compelled to keep on the move. Other examples will be considered from a different point of view in the next chapter, but it must be noted that all these groups were responding to the same problem: it was the rise in the sea level which drove the Doggerlanders away from their hunting grounds into a new land and which provided the Arabian Fish-eaters with a much more abundant source of food than their own ancestors had, and pinned them down to their original homes.

Nearchos' men in their ships repeatedly suffered from hunger, even facing starvation at times, on their voyage; on shore the Fish-eaters were healthy and well fed, and continued fishing. Nearchos and his men seem to have insisted that only wheat or barley bread was a suitable food for civilised men to eat. So they sailed along a shore suffering from starvation where men had made their living from the sea successfully for thousands of years. Alexander's army and sailors refused to eat fish; the Fish-eaters at least were willing to eat bread occasionally.

This was an unusually long-lasting lifestyle, the result of the juxtaposition of a desert land and the abundant sea; but what also helped was the impression gained by people from agricultural areas that the Ichthyophagoi were poverty stricken and on the edge of starvation all the time. So they were left alone by predators (always excepting Alexander's desperate men) and probably also very largely by the tax collectors from the various empires which claimed to rule the area. They had so few material goods they seemed to be hardly taxable; the ignorance of others certainly helped them to survive.

The Gulf and its neighbouring coasts is the one area in the tropical part of the world were a specific and clear human response to the flooding at the end of the Ice Age can be detected and described. In other areas in the hot lands the flooding has left no clear signs of its effects, or of the human response to it. Of course, one may assume that people were drowned, and that others fled before the advancing waters, and no doubt some will also have adapted their lives to the new conditions, but it is only in the Gulf region that the necessary research has been done, so that, combined with the fortunate survival of the accounts of Alexander's journey, we have evidence which has provided a basis upon which to construct a history of the society.

The Bass Strait Islands

Presumably this adaptation happened also in the islands and peninsulas which were the remnants of Sundaland, but the research there is not yet definitive. In Australia, where there were the same problems – flooding and deserts – the issue is overshadowed by the simultaneous desiccation of the interior of the continent (one of the subjects of Chapter 6). One area, however, indicates one possible reaction, or rather alternative reactions.

Tasmania became separated from the mainland of Australia by about 11,000 years ago. The present Bass Strait had been a huge bay until then, and the first separation was near the mainland coast, leaving Tasmania with long peninsulas pointing to the north. As the sea level continued to rise these peninsulas were progressively reduced to the islands which now exist – King Island, Flinders Island, and lots of others. The final separation of all these islands from mainland Tasmania occurred about 8,000 years ago (about the time the Gulf was flooded), but the sea level went on rising for another two or three millennia. So the geographical result was one large island, Tasmania, and a lot of smaller islands littering the Bass Strait. Many of these smaller islands were already inhabited when they became separated from the mainland (Fig. 37).

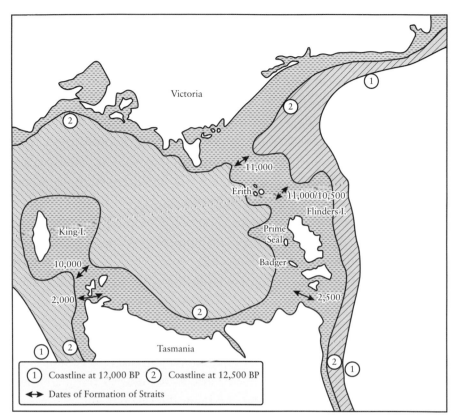

Fig. 37. THE BASS STRAIT ISLANDS. Tasmania and the islands of the strait were attached to the Australian mainland until about 11,000 years ago, but then the oceans flooded the narrow peninsula north of Erith Island. From then on more and more islands were formed, but the decisive breaks came in the next thousand years. By 10,000 years ago, King Island was separated from Tasmania, and by 8,500 years ago Flinders was an island.

Some of the islands, such as Badger Island and Prime Seal Island near Flinders Island, have produced remains dating to the separation period. On the first of these the population died out about 9,000 years ago, at just about the time of its separation; and on Prime Seal Island they lasted perhaps a little longer.

How the end came is not known. Either the people died out or they got away, and either they got away by walking across the final land connection, or they had boats. Within Flinders Island itself, a much larger area of land, the population was apparently also isolated, and had no means of leaving; they died out about 4,000 years ago, having lived on their isolated island for about 4,000 years, with no communication with elsewhere. Yet on King Island, which is about the same size as Flinders and which had also separated off even earlier than the others, the evidence is best interpreted as indicating deliberate abandonment about 7,000 years ago, not long after the time the island was formed; even so, it had had another long preceding period of isolation and separation. The island was later occupied again between 2,000 and 1,000 years ago by people who arrived from Tasmania. They therefore had the capability of reaching the island sixty kilometres across the intervening strait, though they could not, or did not, cross the wider and wilder Bass Strait to reach the Australian mainland. Or at least there is no evidence that they did so – if they even knew that the mainland existed.

The suggested responses range across the possibilities – abandonment by moving to a nearby land when the island was formed, or just before; extinction by dying out; re-occupation after a substantial sea voyage. The differences would suggest that the communities were differently capable of anticipation and planning. The separation of the islands cannot have been sudden; the inhabitants must have been aware of the gradual narrowing of their connection to elsewhere, and it will not have taken much imagination to understand that a final separation was approaching – but perhaps they simply did not believe it would happen.

The King Island people clearly were able to organise themselves to evacuate – and later others (possibly their descendants, acting maybe on a long folk memory) organised themselves to re-occupy the place. The Flinders Island people, on the other hand, seem to have failed to respond in that way, stayed put, and eventually died out; this was the largest of these islands, and they did last for 4,000 years after the island was separated, which implies that food was not the major difficulty; inbreeding in a small original population may well have been the real problem. Tasmania itself, of course, was also isolated, at least from Australia, though it was large enough of itself, and there the people remained and prospered.

The essential difference between the big and small islands was possibly the size of their populations, so long as they were capable of gathering sufficient food. The numbers of people on Prime Seal Island and Badger

Island were probably far too small for survival and continuation. The Flinders Island group were also possibly too small in number to maintain a breeding population, though this is an island of some size, and the reoccupation of King Island indicates that, given the possibility of sailing, and therefore remaining in communication with another population, life on King and Flinders Islands was quite viable. The basic problem was therefore one of the possession of the appropriate technology: the populations with boats have the option of staying – since they can exploit the seas – or of leaving, but those without were doomed to die out.

Denmark

Investigations in Denmark have shown that there was in post-Ice Age Europe a people who had adopted a life which was similar in many ways to that of the Fish-eaters of Gedrosia. Denmark had been on the very edge of the ice sheet. Western Jutland, the North Sea side, had always been free of the ice, and was connected by land with Doggerland and England until the rise of the sea level. It had received great quantities of morainic debris from the action of the glaciers, so that much of Jutland is actually a great moraine several hundred feet thick. Eastern Jutland and the islands, by contrast, had been under the ice sheet, whose leading edge lay north-south along central Jutland; the moraines in the east were smaller, and the scouring action of the ice shows in the intricate geography of the islands and the sea straits.

The proximity of the great sheet of ice, which stretched from northern Germany to the North Pole and beyond, meant that the cold and the ice wholly dominated even the area which was free of ice, and any land close by was effectively frozen even if it was outside the permanent ice cover. Winters, with temperatures permanently well below freezing point, lasted in the ice-free zone for nine or ten months at a minimum, and in some years, even decades, the winter will have been never-ending; any land close to the ice was permanently frozen just below the surface, and that surface was often waterlogged during the summer.

The land was therefore effectively uninhabitable, except for a very short summer period. It was in such warm summers that the reindeer hunters of the Ahrensberg Culture camped at Stellmoor and Meiendorf, a few kilometres south of Jutland, and not far from the ice. They will have known Ice Age Denmark, and hunted over it. But even in summers there will have been few animals available to be hunted in the Danish lands, for the forage was very thin and scarce, and so the animals were few. This was a land which had been crushed and pulverised by the ice in the millennia-long winter. As the ice retreated during the Allerod interstadial – named for the place in Denmark where it was first recognised – and as the water was drained and the sea level rose, Denmark became available for the first time in millennia as a place where new plants and foliage could grow. By about 10,000 years ago the ice had retreated far enough for tundra to form, while the first trees, birch and hazel, colonised as far as the southern part of the Jutland Peninsula. It was colonised, no doubt slowly, by seeds carried in the wind, and by the birds and the few animals; mosses, grasses, and the first trees, birch and hazel, then coniferous trees, and finally deciduous trees, all arrived in succession over several thousands of years.

The warming was dramatic in terms of winter temperatures – a rise of perhaps 20°C in midwinter (though still below freezing point) – but in the summer the rise was about 5°C. It was the frozen nature of the ground during both winter and summer which most retarded or prevented plant growth, and this warming was enough to restart or permit growth. The re-advance of the ice in the Younger Dryas period drove out the new plants once more, but the thawing of the permafrost as a result of the increase in the winter temperatures at the end of that cold period was decisive. By 8,000 years ago oak trees were beginning to grow in Jutland, as well as conifers and limes and alders. Such vegetation provided resources for both men and animals

Along with the forest vegetation there came the forest animals, small deer, rodents, swine, and the forest fruits and vegetables. And the seas warmed. To the east the various early versions of the Baltic Sea formed and drained and re-formed, until the modern situation was more or less established from about 8,000 years ago. The North Sea was flooded by then, and Doggerland was largely vanishing; the present geography of seas and peninsulas and islands existed, give or take some shoreline changes, and subject to the long-term isostatic recovery of the lands which had been under the ice.

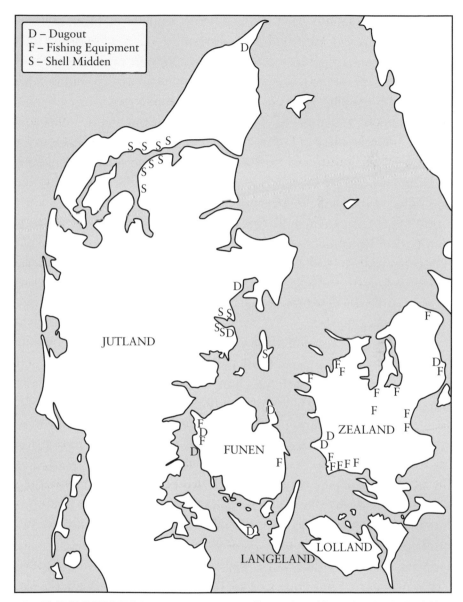

Fig. 38. DANISH MESOLITHIC FISHERMEN. The evidence for the shell midden and fishery societies is displayed, showing the finds of dugouts (including paddles), fishing equipment (leisters, traps, and so on) and shell middens which have been dated to be older than 6,000 years. Shell middens are mainly on the mainland, the other items generally on the islands, though dugouts appear in both regions; nothing is on the North Sea side, which was dry land for much of the Mesolithic period.

This was the land which was colonised by people who, like those who had already moved into Norway, were hunters and fishermen. Their ancestry, of course, had been as the big-game hunters of the Palaeolithic tundra, in particular, perhaps, the people of the Ahrensberg culture, who were already familiar with the area. As the trees advanced they found that their prey – mammoths, reindeer, musk ox, the larger animals of the open tundra – had moved away. Hunting in the forest is much more difficult than in the open tundra, and the animals they hunted there were small and produced less meat. Some of their fellow hunters searched out and followed the greater animals, and moved to the north and east with them. Others, like the families who periodically lived at the campsite at Star Carr in Yorkshire about 10,000 years ago, hunted red deer above all. These hunters by now had domesticated dogs to assist them, and bows and arrows were now being used throughout Europe. These are much more useful in the open forests which had spread throughout the north than are spears and stones.

Those hunters who moved into Denmark found they were in a cul-de-sac, blocked from moving north or east or west thanks to the ice and the flooded Baltic and North Seas. But they could adapt themselves to a different food regime. They exploited both the sea and the shore, living to a large extent, it seems, on shellfish and fish, though their lives were still to a degree migratory, and in parts of the year they apparently moved away from the shore and lived by hunting. They discarded the shells and the fish skeletons into great mounds, called by the archaeologists shell mounds or middens. These accumulated in size over a period of several hundreds of years, which shows that this was a stable and successful way of life in that the same extended families continued to use the same site for many generations (as I have suggested for the campsites in Beringia).

It is necessary at this point to diverge away from Denmark for a space; I shall return to it after this next section.

The Mesolithic Period

The Danish shell middens were built during a period regarded, archaeologically, as coming between the Old Stone Age (the Palaeolithic – the Ice Age) and the New Stone Age (the Neolithic), hence its name is Mesolithic, or Middle Stone Age (though some prefer the term Epipalaeolithic, which seems a spectacularly clumsy, and obscurantist, term). It is a period which is distinctive in more ways than its name. It is distinct from its Palaeolithic predecessor, which is the time of the great big-game hunters of the Ice Age, and from its successor, the Neolithic, which is the time of the farmers. It is particularly characterized above all by the use of microliths, very small prepared and manufactured pieces of stone as components of their tools and weapons ('microblades', in North America). These are very neatly crafted out of flints and other stones, and were presumably brought about by the increasing skill of the stoneworkers, the 'flint knappers', and perhaps by a shortage of suitable raw material in the forested lands.

The Mesolithic is thus the name used for the period in Europe between the retreat of the ice and the arrival of farming, but it can also be applied in other parts of the world. The use of small delicate stone tools and weapons is what distinguishes the Jomon period in Japan, the 'microblade' cultures on the North-west coast of North America – and indeed the Clovis and Folsom cultures in interior North America. It is, as will be seen later, also a characteristic of several of the cultures which existed in India after the global warming which affected that land from about 12,000 years ago. A similar development occurred among Australians. It was, that is to say, a technique which either developed more or less simultaneously in widely separated parts of the world, or spread rapidly through continents and over the seas from a single point of invention. Either scenario seems unlikely, though the former is the better choice, and the phenomenon itself cannot be gainsaid. It is, of course, only one part of a much wider issue, and we shall note it again in the example of the invention or spread of the use of pottery. (And later there were apparently separate inventions of farming in several different parts of the world – see Chapter 7.)

While on the subject it is worth noting that the people living in and about Mesolithic Denmark had acquired or developed other skills besides the manufacture of small and effective stone tools. During the very last period of the Last Glacial Maximum, the first domesticated dogs had become part of man's armoury of hunting equipment, the earliest examples known being found at Star Carr in Yorkshire, as already mentioned. From the Stellmoor site, not far south of Denmark, the earliest known bow has been found. This is an instrument which was, like pottery and microliths, also independently invented elsewhere, in Africa and America, just as the dog was independently domesticated in other regions. And with the extension of the forest of Western Europe dugout canoes became used, this again was an independent development seen elsewhere. It is difficult to say that these were actually invented, though someone must have been the first to go from floating a log to seeing that a hollowed-out log would float better and become more comfortable to use, an idea which probably occurred in several places. These new instruments were decisive in increasing the ability of men to acquire food and dominate the newly-forested landscape. They go along with the increased skill in preparing stone tools to make up the new way of life of the Mesolithic, and possibly were in part a result of that success in such skill. These are not the only cases of the question of repeated invention or the spread of ideas, and it is a matter I shall return to later (in 'Lessons').

Fishermen of Denmark

The new inhabitants of Mesolithic Denmark found themselves living in a relatively small territory, and one which in many ways was dominated by the sea. The rise in the sea level and the flooding of the lowest lands separated part of the land into an archipelago of islands, and the Jutland Peninsula was surrounded by sea on three sides. From the time the ice retreated for several millennia the size and shape of the seas and islands constantly changed, not only because of the ice melting, but also by the isostatic recovery of Scandinavia as the weight of the ice lessened, and by the opening and closing of the connection between the North Sea and the Baltic. In such circumstances one of the necessities of survival was the use of boats, and thus a familiarity with the sea. The proto-Danes of the Mesolithic combined hunting in the new forests with collecting shellfish – which spread north along with the warming of the seas – and with fishing in the seas and straits of the new land. It is the gathering and fishing activities which have left most evidence, because of the middens they built up. In addition, the sea level rise since they lived in Denmark has meant that considerable evidence for their lives has survived under water, and these shallow waters are the ideal areas where underwater archaeologists can operate.

The proto-Danes made extensive use of wood, a material which has usually rotted away on dry land archaeological sites. Their weapons, of course, were mainly of wood (and sinew, in their bows), and their arrows were fitted with microliths for penetration. Similarly, so also was wood used for their houses and their boats. They lived in substantial huts of up to five metres by three, and facing out towards the sea (Fig. 39). These were constructed by first excavating a shallow pit, which was then packed with twigs and leaves to make a slightly raised platform. The packing contained items such as waste flints and the remains of food, indicating that the flint-knapping took place and meals were taken in the same area. Stakes found at intervals around the edges of the platform show that it was roofed; no doubt there were walls of wood or foliage as well. This is very similar in many ways to the huts of the people of Star Carr a couple of millennia earlier.

This is a dwelling which would probably have been used for a single fishing season. It was so flimsy that it would not have survived the winter storms, and the packing of the floor platform would fairly quickly rot and become foul, especially since it seems to have been the receptacle for unwanted items of food, though they were conscious enough of the problem of rotting food so that most of their waste went onto their middens. The food remains show they existed mainly on fish – cod, principally, but also other fish and shellfish – and there were bones of whales and seals; the animals they hunted included red deer, roe deer, and boar – forest animals; and they gathered nuts and fruits. Probably other sites, away from the shore, would show different proportions of food gained by hunting, gathering, and fishing, but it is clear that the fishing was a major part of their lives, but not quite so exclusively as among their contemporaries in the Gulf and Gedrosia.

The waterlogging of their homes has also preserved evidence of their fishing methods. They used bone hooks, wooden traps, and nets, so much of their work was done from close to the shore. Several examples of their dugout canoes, or parts of them (including their paddles) have been found. These date back 7,000 years and more (Fig. 40). At least six of

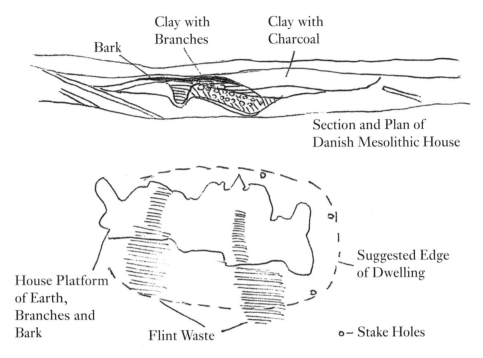

Fig. 39. A DANISH MESOLITHIC HOUSE. This shows the remains of a house discovered in an underwater excavation at Mollegabet, of the island of Langeland. The section shows the stratigraphy of the site. The house was formed of a layer of bark covering the earth and branches platform, which raised the floor above the surrounding land by a few inches. The covering would have been of bark and branches, supported by the stakes, whose holes are visible on the right in the plan. The flint waste indicates part of the work being done; scattered around were flints, bones, antlers, signs of burning. It is unlikely that this site was used for more than a single summer season.

these canoes have been found along the Danish coast of that approximate date, often in connection with other fishing gear.

They also built fish traps. They constructed fishing fences in shallow waters, which were used to guide the fish into areas where they could be easily caught and speared – or kept alive until required. They even, it seems, used lights to fish at night: one dugout was found containing a clay patch on which a fire had been burnt. The interpretation is that this fire was used at night to attract fish (a practice which has now graduated to the use of large arc lamps in other parts of the world); the alternative is that the canoe was being used to transfer fire to a new residence. The combination of dugout canoes, fish traps, spears, and hooks, makes for an obviously well-thought-out campaign of fishing. The other main part of their fishing activities was to gather shellfish, notably oysters, but also periwinkles, whelks, and cockles. These were probably used both as food and as bait for catching swimming fish. The shells of these animals went to form and increase the great shell middens, but they were placed at some distance from the houses, no doubt because of the stench of rotting fish they would produce.

The remains in Denmark indicate a well-organised and efficient system of food collection, using a variety of methods, all of which were presumably successful, since they continued in use of several thousands of years – indeed, most of them are still used to this day. These proto-Danes were not wholly reliant on one type of food, however, for they hunted on land as well as fished. They were mobile both on the water, using dugouts, and on land, for, like their counterparts in Oronsay in the Hebrides (see Chapter 5), they moved with the seasons, exploiting different ecologies – hence the temporary nature of their huts. It was clearly as successful an adaptation to the new climatic conditions of the places they lived in as that of the Gulf Fish-eaters; indeed, it was perhaps more successful, since it was less dependent on a single food – which was, of course, a result of living in a more benign region. The importance of the sea and their boats is emphasised by the discovery that several of the thirty or more Mesolithic boats which have been found in various parts of Europe were used as burials (Fig. 40). Only a community intimately attached to using the sea would do this.

This society originated at about the time, 9,000 years ago, that the decisive flooding of the North Sea took place, about the time when the new Norwegians had arrived and settled, and at about the time the Gulf fisherman faced the flooding of their old home and their need to adopt sea fishing. It was a society, like the Fish-eaters, and like some other Mesolithic societies, which was stable and which lasted a long time, and was therefore successful. In Denmark its later manifestation is the Ertebolle culture, with much the same characteristics as the preceding society of fishermen, but situated on the southern Swedish coasts, and together these succeeding cultures lasted until about 6,000 years ago, being overtaken eventually by the arrival of farming.

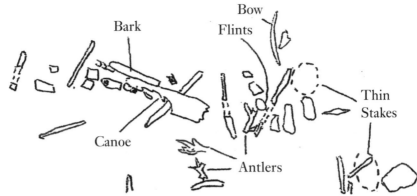

Bark   Bow   Flints   Thin Stakes   Canoe   Antlers

Human Bones

Fig. 40. A CANOE BURIAL. This is a plan of the remains of a burial, now under water at Mollegabet, Langeland. The remains have been disturbed by passing ships, but enough was left to show bones of a young man lying on the remains of a wooden canoe. The sheet of bark may have been used to wrap the body. He was buried with offerings of reindeer antlers and a bow, showing he was mainly renowned as a hunter. The long stakes may have been used to support the canoe above ground, or possibly to form a shelter over it.

Comparisons

Looking at the Danish fishing communities of the Mesolithic, the Fish-eaters of the Gulf, and the stranded Australians on the Bass Strait islands provides us with a comparison between similar human adaptations to very different climates and geographies. The Bass Strait Islanders had the most difficult task, in that their islands were evidently too small to support them over the long haul. The Gulf people clearly had a tough time as well, since their land resources were very limited once the relatively brief humid period had ended. They lived on a desert coast, so their options were very few. In that sense, unlike the Danes, who could still exploit the resources of the forest, they were largely driven to exploit the sea, since that was the only source of food they had, apart from palm trees. Yet they did succeed in establishing a competent style of life which their descendants followed for millennia. Mesolithic Denmark, of course, was much more prolific in land-based foods – huntable animals and gatherable fruits and nuts – but for several thousands of years the seas and the shores in the Danish islands were exploited with some intensity as a preferred option by the people who lived there. The Bass Strait islanders, without the capability of making boats, could not exploit the sea, perhaps also because they could not make nets and lines.

The two successful styles of life, in Denmark and in the Gulf, developed more or less simultaneously, and while they continued successfully in a stable and hardly changing lifestyle, the Bass Strait islanders failed. The earliest dates for the Danish Mesolithic are in the eighth millennium BC – 9,000 years ago – though most are between 8,000 and 4,000 years ago. This was a little earlier than the earliest dates for the Gulf people, who, of course, could not develop their skills until the Gulf flooded, whereas the proto-Danes moved in soon after the ice retreated. These three groups are so far apart geographically, yet so nearly contemporary in time, that they clearly represent independent adaptations to the new conditions which were imposed on them by the end of the Ice Age. One of these 'adaptations', if the word may be permitted, was to surrender to the difficulties, and either leave (as the King Islanders left) or die, as did those on other Bass Strait islands. Those who left did so in order to continue their old life in another location (so they should perhaps be included with those noted in Chapter 1); dying in the face of overwhelming adversity is also, of course, an option, and one which some societies undoubtedly adopted. The Flinders Islanders did struggle on for a long time, dying at about the time the proto-Danes turned to farming.

The crucial element in success or failure was the ability to exploit the sea. The proto-Danes and the Fish-eaters both developed their sea fishing skills at the same time, and this implies strongly that other humans elsewhere were thinking out ways of coping with the new world and were coming to similar conclusions; the Bass Strait islanders did not apparently think the sea was an option they could adopt.

Responses to Flooding

In the cold regions the human response to the flooding was, so far as can be judged, rather different from that in the more benign areas. In the North Sea the threatened hunters scattered, to Norway, to Britain, to the continent. In Beringia there were successive but contradictory moves; one movement was southwards, at first into the maritime landscape of the North-west coast which also compelled a reliance on fishing, but the possibilities and resources of that coast were much greater than in the Gulf, and other people (perhaps later arrivals who found the limited number of places already occupied) moved on to the rest of North America; those who stayed in the North-west islands and peninsulas could exploit both the sea and the land, and the land was rich in forests and huntable animals, and fish in the rivers. In this they were similar to the proto-Danes, and to the Gulf Fish-eaters, in their land of fewer resources; both successfully adapted to the possibilities in their new lands. The other reaction was that of the people who remained in Alaska, the Aleut and the Inuit, who adapted their lives to the new local situation, shifting to fishing and to inhabiting the remaining ice-lands. In Japan the response to the flooding was again different, as will be discussed in the next chapter, but it certainly did not involve a long-term millennia-long complete reliance on coastal fishing.

The response of the fishermen of the Gulf and the Gedrosian coast was, however, unique, though it can be compared with the Danes and others. Their land was unique in its problems and its resources, or lack thereof. Their original response, in the Gulf itself, was to resort to fishing, and they were able to expand into trading later, an adaptation their successors continued. On the Gedrosian coast the Fish-eaters did not have the trading option to expand their resources, though they took advantage of the riches available from the sea as a compensation for the absence of resources on the land. It was a pared-down version of the Gulf fisherman, whose location was clearly somewhat more favourable. To some degree their lifestyle was forced on them by the limits they faced, but they still clearly had the element of choice. They could have simply walked away and become hunters elsewhere, say in interior Iran, a relatively short journey, and one which they clearly did take at times, for this was where their items of iron and pottery came from. In Babylonia the rivers still ran, and animals could be hunted – indeed no doubt some of the people who had lived in the unflooded areas will have taken up those options. But the Gulf people and those of the Gedrosian coast took the opportunity presented by the new sea and did exploit it to provide themselves with a satisfactory life, one which allowed their numbers to increase, for there were certainly more people living along those shores than there had been in the area before it was flooded. (One of the marks of a successful style of life is its ability to support a larger population than before.) Like those who moved to Norway, the Fish-eaters of the Gulf showed that they were capable of understanding and exploiting the new conditions, adapting to them, and turning them to their advantage over the long term – several thousands of years, in fact.

The Bass Strait people, however, showed what happened without a deliberate adaptation: extinction.

Further Reading

The archaeological investigation the Fish-eaters is detailed in Mark J. Beech, *In the land of the Ichthyophagi, Modelling Fish Exploitation in the Arabian Gulf and Gulf of Oman from the 5th Millennium* BC *to the late Islamic Period*, British Archaeological Reports S1217, Oxford 2004; the two ancient sources are Strabo, *Geography* 16.2, and Arrian, *Indike*, 26–31.

For Mesolithic Denmark see David J. Quill Smart, *Later Mesolithic Fishing Strategies and Practices in Denmark*, British Archaeological Reports S111, Oxford 2003, and I.B. Enghoff, 'Fishing in Denmark during the Mesolithic Period', in Fischer (ed.), *Man and the Sea* (see Chapter 2); for the house and the canoe burials see Jorgen Skaarup and Ole Gron, *Mollegabet II, a Submerged Mesolithic Settlement in Southern Denmark*, British Archaeological reports S1328 Oxford 2004. There is also a brief report on underwater archaeological discoveries in Denmark in *Current World Archaeology* 17, July 2006 ('Sunken Stone Age').

For Tasmania see Nick Porch and Jim Allen, 'Tasmanian Archaeology and Palearchaeological Technological Perspectives', in *Antiquity Transitions* issue, 714–732, and Flood, *Archaeology of the Dreamtime*, Chapters 9 and 13 (noted in Chapter 1).

Chapter 5

Sedentary Foraging

Both the Gulf/Gedrosian Fish-eaters and the proto-Danish fisher-hunters can be classified as Mesolithic people who hunted for a living, either at sea or by land, but they are different from the normal hunters in that they stayed in one place; in this they are similar to many of the groups who will appear in this chapter; they were singled out to be treated separately to make the point about the adaptations possible as a result of the changes in the sea level. This new chapter deals with another aspect.

There are others in their category who were not so concerned with hunting and fishing for their living. They did not ignore such a resource, any more than we, their descendants, ignore it, but it became a less important part of their lives. These were people who, as a result of the change in the climate in the several parts of the world at the end of the Ice Age – not just the level of the sea – were able to adopt a style of life which was much less nomadic and more settled. Apart from the Gulf Ichthyophagoi and the Danish midden-people, the new inhabitants of the North-west American coast and islands had adopted a similar semi-settled life, living partly by hunting and fishing, and partly by gathering land foods and seashore molluscs. But the groups whom I wish to consider in more detail here went further: they became sedentary or near-sedentary foragers.

This is a term being now being used to classify groups of people who collected their foods from a particular and usually well-defined and limited area. They were settled more or less permanently in one place, or perhaps a few places fairly close to each other, staying there for a fairly lengthy period. From their village they moved out to gather the foods and other resources they needed, like the proto-Danes who made fish traps and built shell middens. This was such a successful lifestyle that they have even been termed 'affluent foragers', which is perhaps going too far. Four groups of such foragers in particular have been studied in some detail: in Japan, in the Scottish Hebrides, in India, and in South America.

Japan

The peninsula which became Japan remained attached to the continental mainland throughout most of the final phase of the Ice Age, the Last Glacial Maximum, when the sea level was at its lowest, and the ice was at his most extensive. The Sea of Japan was at times a lake and others had one or more outlets to the ocean, but crossing to Japan from Korea was never difficult in most periods and at times was without any impediment (Fig. 33). During that time the land was thinly occupied by well scattered hunter-gathering groups, who made and used crude stone tools typical of the Palaeolithic everywhere, and who moved regularly to new campsites; they were therefore similar to every other group in other lands who lived close to the ice. No doubt they also moved in and out of the area which became Japan.

There was one unusual element in the lives of these people, however. The island of Hokkaido in the north appears to have been separated from Honshu, the main island, for long periods late in the Ice Age. Indeed, at first, Hokkaido was itself divided into two parts. Only in the Last Glacial Maximum, when the sea level had been lowered even more than before, were Honshu and these northern Hokkaidos all joined into one. Yet even before that junction people had reached Hokkaido, and they could only have done so by boat. Elsewhere it is clear from the findings of obsidian at mainland sites that travel by sea was reasonably common along the Japanese coast even at the end of the Ice Age, for the obsidian can only have been collected from the island of Kizojima fifty kilometres offshore. This was as major a voyage over the open sea as anything the new Norwegians, or the settlers moving along the north-western American coast accomplished. It argues a high level of skill in seamanship and in boat-building, as early as the end of the Ice Age; it is perhaps not surprising that the Japanese boat builders lived not too far from the Aleuts, who also built boats, though the materials they used differed – wood in Japan, skins by the Aleuts – which was no doubt the result of the differences in their available resources, and there is no indication that the two were in contact.

During the final Ice Age phase, down to about 12,000 years ago, the Japanese peninsula became separated from the mainland of Korea by a new strait, the Strait of Tsushima; the narrow straits in the north between Hokkaido and Honshu, Hokkaido and Sakhalin, and others, were flooded. The most important was the Strait of Tsushima, which was, at first, only a narrow channel, through which the otherwise enclosed Sea of Japan drained into the ocean, and which widened as the sea level rose. The inhabitants of the future Japan were not actually seriously cut off from the mainland until the sea level rose decisively after the sudden rise associated with the '8.2K Event' in North America – but even then their possession of boats meant that connections with the continent continued. Japan was never wholly separated from the continent any more than was Britain, except geographically.

(It may be pointed out that these boats, while generally dugouts – Japan was never substantially deforested, unlike Ice Age Europe – indicate a much more prevalent familiarity with the sea than is normally assumed at this time. This clearly has a relevance for the North Sea migration to Norway, and even more for the theory of the coastlands' colonisation of North-west North America. From Japan to Kamchatka to the Aleutian Islands – a string of islands and peninsulas the whole way, all of them inter-visible – is by no means an unimaginable journey, or series of journeys, even in a dugout; the use of dugouts in Japan at this very early time, their appearance as soon as forests grew in continental Europe, and their use in Denmark, all suggest a similar response to the new conditions which came at the end of the Ice Age among a wide variety of peoples. The Pacific Ocean was clearly ringed by a series of societies with maritime capability, by the end of the Ice Age.)

The Tsushima Strait between Japan and Korea widened drastically from about 10,000 years ago. Not only that, but the long single island of 'Japan' became an archipelago as the sea flooded the low-lying valleys and separated the large island into several. During these events Hokkaido and Honshu were separated again, and the southern islands of Kyushu and Shikoku also became separated from Honshu. This rise in the sea level rose to a higher level than at present, and many of the lowest valleys, which are now dry, were flooded for a time, so that many of the archaeological sites in this period are uniformly higher up the valleys or the valley sides than later sites. Developments which appeared about that time spread widely throughout the various sections of the country, and this makes it clear that the geographical separations were not a matter also of social isolation – that is, voyaging across the Strait of Tsushima between Korea and Japan and between the several Japanese islands probably never ceased.

As the Last Glacial Maximum relaxed its fierce grip and the sea level began to rise, two successive developments took place, both of which contributed to the next stage in Japanese society's history. By about 14,000 years ago, still within that Last Glacial Maximum, stone tools being manufactured became smaller and neater, and more precise – they were microliths, that is. On the American North-west coast these are called microblades, and in Europe this development is characteristic of the Mesolithic period, which follows the retreat of the ice, and it also takes place in India, as we shall see later in this chapter. The development in Japan was slow, as it was elsewhere. The new method was to use the small neatly-shaped flints and other sharp stones (such as obsidian) by setting them into wooden hafts or onto the points of arrows, so producing a sharper edge or a better penetration than the larger stones and sharpened or fire-hardened wood of earlier millennia. (Bows seem to have been in use in Japan almost as early as in Europe.) But the earlier tools were not forgotten; the result was an increase in the quality and variety of the tools available, and an expansion in the capability of the hunters. By 12,000 years ago tools using microliths were being manufactured in most of Japan, and even in the more isolated northern Hokkaido soon after that.

The other development arrived somewhat later, and certainly came from the continent. The microliths seem to have been a local development without outside influence, though this is not certain; as usual, either notion – independent invention or imported knowledge – causes problems of understanding; similarly, the bows may have been a local invention, but these could well have arrived from the continent as well. The most startling new development was the manufacture and use of pottery, and this certainly seems to have arrived from the continent. The earliest known pottery (other than the accidental baking of clay, or its use in hearths, or for lining basketwork) was made in eastern Siberia, in the Amur Valley region, and its use reached Japan about 14,000 years ago, and so just before the end of the Ice Age.

The concept of, and methods of making, pottery were probably therefore brought into Japan by migrants from the continent – DNA investigations have shown that the basic Japanese population came originally by way of Korea. It is, of course, an important addition to any society's toolkit to have the use of pots of a variety of types and sizes – and once invented or introduced all sorts of different shapes and forms and sizes can be made. Quality was usually fairly low to begin with, since the origin of pottery seems to have come from the practice of using clay to line baskets which had been made by plaiting flexible branches, thus making the baskets able to hold water. Once heat was applied, it was quickly realised that the wooden framework was not necessary, and that pots could be made by hand. Further, the smooth outer face of a pot was always liable to attract decoration, for the softness of the clay before firing was apparently an irresistible medium for doodling, with fingers, or with paint. Here was yet another outlet for the human propensity towards art (Fig. 41).

The separation of the Japanese peninsula from the continent came sometime after the introduction of pottery. It took place about 10,000 years ago, and Japanese society then developed along lines already marked by the increased sophistication and technical artistry of both the microlith stone tools and its decorated pottery. This is called the Jomon period of Japanese archaeology, and in its successive phases it lasted 8,000 years, finally giving way to the Yayoi period, in which agriculture became dominant, during the last millennium BC. The maritime aspect was retained also, and sailors continued to utilise the obsidian source on Kizojima. This mastery of the local seas meant that fishing could also be a major source of food. The people of Hokkaido and Honshu were thus always in communication, and contact was maintained with the Korean continent. Japanese isolation, like Britain's, was always only relative (Fig. 42).

The new Japan of the post-glacial period, surrounded by the sea, warmer than before, and ice-free, and with forests spreading in the mountains, was a place where hunters had only fairly small animals for their prey – the larger animals largely failed to survive the climatic change,

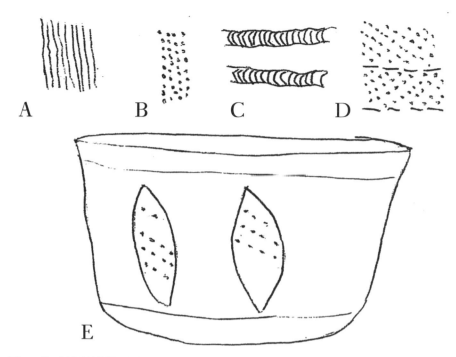

Fig. 41. JAPANESE POTTERY DECORATION. From the very start Japanese potters used various means to decorate their products. These are all examples of 'Initial Jomon' pots (before 8,000 years ago), the oldest known:
A. Incised vertical lines; B. cord impressions; both of these are from pots found at Natsushima in Tokyo Bay; C. Nail impressions – this pot had cord impression work on its lower half; D. Diagonal cord impressions; these are from a pot from Kitasharkowa, Kyoto. E. A bowl from Natsushima decorated with bold incised lines, and with cord impressions in the ovals.

either being trapped and killed off by the greater warmth and the change of vegetation, or being quickly hunted out. Boar and deer were now the largest animals (just as they tended to be in Europe once the mammoths had gone). On the other hand, the land was relatively rich in vegetable and fruit products, and the seas had fish and sea mammals in abundance, many of which were catchable on or near the shore, and the shores had plenty of shellfish. The geography of Japan, with a long mountainous spine flanked by lowlands along the coast, meant that the sea was within relatively easy reach of almost every inland region, and this encouraged the use and exploitation of the sea. These resources were the bases for a society which was much less dependent on hunting by land than the

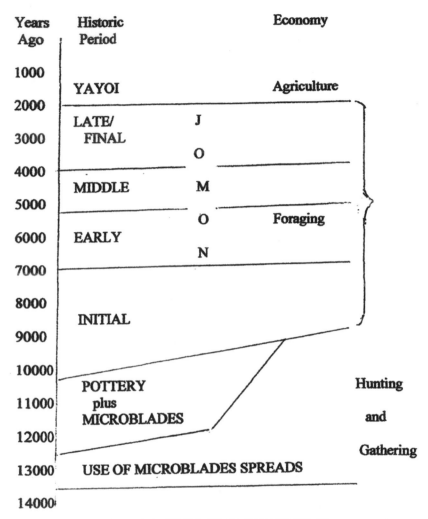

Fig. 42. THE PERIODS OF JAPANESE PREHISTORY. The very early use of pottery distinguished Japan from most other areas, as does the very late adoption of agriculture. So the German period – 8,000 years long – is that time in which foraging was the dominant means of gaining a livelihood.

previous hunter-gatherers. Instead it became self-sufficient on the basis of some hunting, where possible, but mainly on foraging for food on land, and on fishing by sea – that is, utilising all available sources of food. Also, due to the geography, it became possible to conduct all these activities from a fixed base, initiating sedentary occupation (Fig. 34).

It has been possible to develop a theory, at least in general and rather idealistic terms, of the general life of the subsistence foragers in Jomon Japan, based on the notion that they lived through a regular annual cycle. Winter in much of Japan is cold, and at that time the vegetable foods are less available. This would be the time for hunting, the prey being the larger animals which do not hibernate, such as deer and boar. It was at this time that houses were built and repaired, and, especially if the weather prevented or inhibited movement, for the manufacture of new stone tools (though this surely took place at all seasons). Fishing was also less productive in the winter, and with the seas more dangerous then there would be long periods when sailing was not possible, especially in the fairly primitive boats available, but it increased in importance from the spring onwards, so that summer and autumn saw the predominance of hunting for sea mammals and mollusc-gathering, both activities conducted along the shores, as well as fishing in rivers and for the sea fish. Spring and autumn were the main periods for gathering fruits and vegetables and tubers, much of which could be stored to help out in the winter.

The evidence for much of this, which is not very surprising once it is understood that the settlements involved did little hunting and no agriculture, is in the excavations of the shell middens which line the early post-glacial shorelines which are somewhat inland from the present shore (again note the similarity with Denmark), the result of the temporarily higher sea level (figs. 34 and 43). These have also produced evidence of the tools and weapons being used – the latter included a sophisticated harpoon, and dugout canoes – and the burials of dogs and humans. Dogs were apparently as carefully and deliberately buried as the human dead, implying that domestication had taken place (almost as early as at Star Carr – yet again a parallel but separate, indigenous development); the humans were generally about five feet tall and slightly built – very similar to the build of much of the present Japanese population.

The point to be emphasised here is that this style of life brought people into the practice of exploiting a particular area of the country. The big-game hunters of the Palaeolithic had necessarily ranged over wide areas, because their prey animals were migratory, and so the hunters were having to move constantly in order to find food. But by concentrating –

of necessity – on smaller game animals, vegetable foods, and fishing, the people of the permanently settled and occupied Jomon villages in post-glacial Japan ranged over much smaller territories, but still succeeded in gathering a wider variety of foods. In many ways Japan was an ideal place for such a life, for a community could be based in a coastal plain where they could have access to the interior mountains, to the rivers and forests, to the sea and the seashore, all within a relatively short distance.

The particular scheme outlined above of a pattern of foraging has been suggested as the seasonal lifestyle in the Jomon period, but it is also a very general sketch of one, and takes little account of the varieties of landscapes in Japan, which tends to mountains with many relatively small coastal plains, nor of the differing climatic regimes, which range from sub-tropical in southern Kyushu to cool temperate in Hokkaido. The mountains are forested, the coastal plains could be marshy, so the subsistence regime suggested by the scheme of foraging must be applied with discrimination, taking into account the different climatic and landscape regimes.

This is, in fact, one of the main points of the Jomon period, and of the Mesolithic generally in other regions: that once people settled into a defined area, they developed their own styles of life and their culture away from their neighbours – quite apart from having a decisive effect on the land they were exploiting, something which had not happened in the Palaeolithic. This is the case in Japan, just as it is in the Gulf or in Denmark. And yet they were not wholly isolated (any more than the hunting bands of the Paleolithic were isolated from each other). They also communicated with their neighbours, exchanging each other's characteristic products, exploiting unique resources and transferring their products onwards, perhaps by trade, perhaps by exchange of gifts. In the process of this interaction they heard of, and copied, each other's achievements. In Jomon Japan this can be detected even in the 'early Jomon' period between 12,000 and 10,000 years ago – and, of course, even before then with the exploitation of obsidian sources, which is a form of trading.

Without the need to hunt animals, except perhaps in the winter, and perhaps only occasionally, it ceased to be necessary for people to migrate, though mountainous Japan is not a particularly suitable country

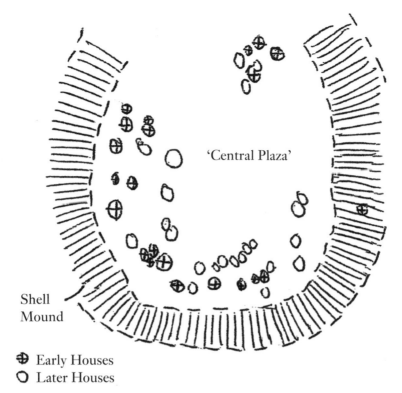

'Central Plaza'

Shell
Mound

⊕ Early Houses
◐ Later Houses

Fig. 43. JAPANESE SHELLMOUND VILLAGE. The shell mound at Takari-kido in the Canto region of Japan is of the Middle Jomon period mainly. It was occupied for a long time, during which the shells were discarded all round the edge of the village. The houses were clearly constructed to plan, leaving the 'central plaza' clear, and an opening to the north. Four phases of houses were diagnosed by the excavators, based on pottery styles; here they are consolidated into two; several other houses were also found, but were undated.

for a migrant society anyway. It thus became possible to live more or less permanently in one place. One is always close to the sea, to a river valley, to the mountains and forests, and so to all the resources available in all those habitats, when in Japan. Resources of food, wood, fish, forest products, stone, clay, were all usually within fairly easy reach of most villages. As in Mesolithic Denmark, relatively permanent houses could be built and inhabited for lengthy periods (Fig. 43).

As the Jomon period passed through the stages which the archaeologists call 'Initial', 'Early', 'Middle', 'Late', and 'Final', this foraging way of life which had been developed in the wake of the Ice Age was maintained and spread to all the islands, modified somewhat as it spread by climate and geography. It was eventually supplemented, and then replaced, by rice agriculture, but not until the first millennium BC. Sedentary foraging in Japan was, that is, a way of life which lasted for perhaps eight or more millennia. It was clearly another successful adaptation to the new conditions imposed by the warming of the end of the Ice Age.

The Earliest Pottery

One of the developments which drew particular attention to the Jomon period was the discovery that pottery manufacture had been pursued from a very early time, and in fact, the earliest examples of crude pottery were made in the period before the Jomon, in the Japanese Palaeolithic, about 14,000 years ago. The earliest examples in Japan are from the southern island, Kyushu, at the site of Fukui, though other early examples have also been found in Honshu and Shikoku – that is, pottery was an established manufacture throughout western Japan even before the Ice Age finally released its grip on the land, and even before the Mesolithic began.

Japan was only one of a number of areas in the Far East where pottery was made at this early period. It was manufactured in several parts of China, particularly in the eastern interior, the modern provinces from Manchuria to Shanxi and south through Hopei and into Guangxi – a huge great swathe of country in which at least ten separate 'complexes' – that is, societies – which made use of pottery in the earliest years have been distinguished. There was also another nearby region, centred on the Amur River of the Russian Far East province, which seems on present evidence to have been the earliest area of pottery manufacture of all.

It is difficult to see all this development, relatively closely concentrated as it was, as a series of independent inventions. In this case it is perhaps easiest to imagine a single area of origin of the methods of manufacture, with improvements and elaborations by others incorporated in the practices as the knowledge of pottery manufacture spread. Making pottery is, after all, not a particularly arcane process, but nor is it simple. It demands an understanding of clay, control of fire, and of the process of firing, but all this can be fairly easily grasped and improved on once the basic procedure is understood. However, the speed with which the use of this new tool was adopted throughout a large area is significant. Once again this is a response by wide variety of peoples and communities to the problem posed by the end of the Ice Age and the decline of the wider nomadic hunter-gathering lifestyle, though in this case it seems that a single originating centre can be distinguished.

From the Russian Far East to western Japan to eastern and south-eastern China pottery came into use, and this was later the area where a more sedentary foraging lifestyle of the Mesolithic period was articulated. The unlikeliness of independent inventions means that we must envisage the transfer of the technology from one social group to another quickly, and this is obviously more likely than separate invention. The combination of the two developments at much the same time is also likely to indicate that they were connected. The decline of the Palaeolithic nomadic and hunting way of life did not stop people from moving, and connections with other societies were constant, if perhaps not necessarily frequent. The use of ceramic containers both assisted in the preservation and preparation of food, and at the same time made it more difficulty to keep moving, since the pots themselves tended to be heavy and awkward to move about. (Incidentally, there is also evidence of ceramic use in parts of Mesolithic Europe, but not before about 8000 years ago – time enough, perhaps, for the knowledge to have percolated westwards. But the earliest examples are from Denmark, with, as yet, no examples from Eastern or Southern Europe so early – independent invention is here certainly possible. It is of interest that it was in Denmark, where the shell midden culture was a similarly foraging one, like that in Japan. The use of pottery is thus something which seems to develop as result of adopting a sedentary lifestyle: independent invention, but one which developed out of similar conditions.)

The possession and use of pottery vessels was not actually necessary for successful foraging, but it was understandably a useful accessory as well as imposing a certain limitation. Indeed, if it was necessary to move regularly, as the Danish hunter-fishers did, the ownership of fragile and bulky and awkward pots might be a positive hindrance. The pots are used most importantly for the purpose of holding liquids, and, when they were tough enough, for cooking, though they could also be used for storing foods, and this would be a useful factor for foragers, who could thus preserve foods from the harvest season onwards. (The Palaeolithic site of Dolni Vestonice in the Czech Republic has produced clay ovens, or at least cooking areas made of clay, but this is not the same as making portable pottery of clay.) Vessels made of leather could not be put to the same use, though these were lighter in weight, more flexible, and less fragile, and so easier to transport, and this was no doubt what was used by the mobile hunters, but carrying liquids in them was barely practical. So the absence of pottery is not an indication of backwardness, only a possible indication of a style of life in which it had no useful place. Yet the manufacture of pottery is a clearly a technological development of some importance, indicating as it does an increased mastery and control of fire and heat, and a wider and more detailed exploitation of natural resources.

Pottery was not, therefore, an essential item of equipment for people of the Mesolithic. There were other places than the Far East where more or less permanent settlement took place in that period, but where pottery was not present, though where such indications as the presence of shell midden mounds show long and/or repeated occupation. The essential point to note is not the variety of materials used, or the foods eaten, or the country occupied, but the fact that the Jomon people were almost wholly sedentary, occupying a fairly well-defined region of land and exploiting it with care and close attention to its products. And there were other countries which use the same system.

Oronsay

A variation on this theme of foraging is visible in a series of camp sites which have been found and excavated on the Scottish island of Oronsay. Oronsay is a small T-shaped island, less than four kilometres on its longest axis, in the Inner Hebrides, seasonally connected to Colonsay next door by an almost-fordable channel, only 200 or 300 metres away. (The island of Oronsay was in fact even smaller in the Mesolithic than it is now by perhaps thirty per cent, as the sea level was then at its highest, and the isostatic recovery had not yet had its full effect.) This is now a very remote spot – though that depends on one's original viewpoint – but it seems that Scotland was populated after the Ice Age first by people arriving by sea along the western seaboard, and so by people who sailed

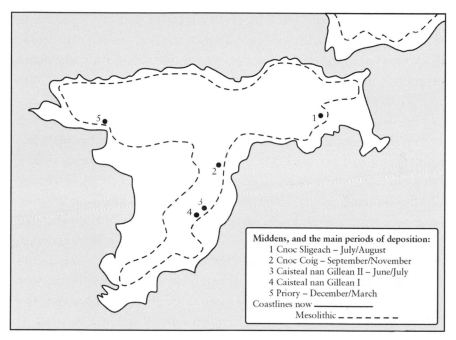

Fig. 44. THE FORAGERS OF ORONSAY. The island of Oronsay was considerably smaller in the Mesolithic than it is now, and the middens are all close to the Mesolithic shoreline. (It also meant that the island was more difficult to reach, necessitating the use of a boat.) The circuit of the island is indicated by the numbers and the seasons, based on the remains of staithe in the middens.

among the islands and along the coast as readily as they travelled by land (Fig. 44).

The ice cap on the Scottish Highlands finally melted away by about 13,000 years ago, though the later cold spell of the Younger Dryas may have deterred settlers, even if it did not fully restore the ice sheet. Isolated finds of small arrowheads suggest that hunters roamed over the country after the ice had gone, but the Highlands will have taken a long time to develop an ecology favourable to animal life: trees, for example, do not seem to have been able to grow there until well after the Younger Dryas cold spell ended, and so after about 10,000 years ago; the earliest Mesolithic site is dated to not long after that time. This was at Kinloch on the island of Rhum about 9,000 years ago – Rhum is farther north than Oronsay, but is just as isolated. (The gap between the end of the ice and the arrival of the 'first men' is often regarded as a puzzle, without considering the actual conditions at the time: one popular explanation is to suppose that the earlier evidence is under the sea, covered when the sea level rose – an all too easy excuse; in fact, for quite some time – centuries, probably – the country will have been barely habitable by humans in the absence of useful vegetation and animals to hunt.)

The Mesolithic evidence for Oronsay is dated to about 6,000 years ago and after, that is, after the end of the Younger Dryas cold period. So it is not the earliest occupation in the region (that on Rhum being earlier), but it has been interpreted in a very interesting way. A series of five shell and fish middens of the Mesolithic period at several places on the island have been investigated. Each was situated on what had been the shore line at the time: on the east is Cnoc Sligeach, and on the west, facing the Atlantic, the Priory Midden; down the southward pointing shank of the 'T' of the island are Cnoc Coig and two other middens adjacent to each other, Casteal-nan-Gillean I and II. Two of these middens are very large indeed: Casteal-nan-Gillean II is 30 metres round and over three metres in depth, and Cnoc Sligeach is only slightly smaller.

All of the middens have produced slightly differing evidence of human occupation, but they were all composed of bones, probably of seals and otters, and shells of molluscs, with some traces of red deer and pigs. The mounds were, of course, in use over a very long period, and probably intermittently. The layers of remains were interleaved with layers of blown sand, which might imply periods of non-use and exposure to the elements (or, of course, the deposition of the sand could have been produced by a particularly violent storm on a single day, or several storms, a not uncommon event in the region). Occupation evidence included hearths and post holes, probably for huts and tents.

Investigation of the contents of the middens revealed interesting anomalies. The bones of deer and pig which were found, for example, were not those which had held meat. The deer bones were mainly the antlers, and so it is assumed that they were brought from elsewhere to be made into tools (though why the tools were not made elsewhere so that the awkward antlers did not need to be transported is not addressed). Neither deer nor pigs seem to have inhabited the island at the time.

For the seals and other marine animals, however, all the bones of the body were present, so these clearly formed the main diet of the midden builders, and were consumed on the spot. This is a conclusion supported by analyses of some of the human bones found in the midden, which indicated that meat such as that of seals was consumed to the exclusion of almost anything else. Both of these human remains came from one midden, at Cnoc Coig, and the seal remains in that midden included three pups, which suggests that this was the probable base for the seal hunters in the autumn, when the seals were new-born.

However, it is by an examination of the ear bones of the staithe fish that the theory of seasonal occupation of each midden is derived. All staithes are born at the same time, and grow quickly and at a steady rate, so the size of the ear bones indicates when each individual fish died. The findings show that Casteal-nan-Gillean II was the midden in use in the early summer (June and July) when the staithe bones were discarded; in the midsummer Cnoc Sligeach was in use (July and August); Cnock Coig was in use in autumn (September to November); and the Priory midden was in use in the winter months. This was not a ritual cycle of movement set in stone, of course, and Cnoc Coig saw another period of occupation – or perhaps deposition would be a better term – in the spring and Cnoc Sligeach in the winter, while there are also some signs of the continuous use of all the middens throughout the year – but the pattern is nevertheless very suggestive. (Though the winter occupation of the site most exposed to the winter Atlantic storms, the Priory Midden, is curious.)

From all this it is theorized that in the Mesolithic period a group of people could have lived on the island all year round, moving with the seasons from one midden site to another. Alternatively, from the finds of red deer and pig bones, it would seem that the early inhabitants, or possibly a later group, had contact with other islands and with the mainland – which is obviously highly likely given the inhabitants' likely familiarity with the sea – how else did they get to the island? – so that the middens on Oronsay might be the result of intermittent occupation by temporary visitors at particular times of the year. (The island is easily accessible today from its neighbour Colonsay, even without a boat, though this would be more difficult in the Mesolithic, when the strait between them was wider; we know, of course, that the people had boats then, since that is how they must have reached both Oronsay and Rhum in the first place.) As a third alternative it could be that the middens were accumulated by different population groups, all using the island at the same time. The island was, however, at that time scarcely large enough for more than one foraging group at a time, and the finds do imply that the remains were discarded at different times of the year, implying a regular cycle by a group of people familiar with the island and its resources.

None of this is definitive, and the most attractive theory – the circulation of a group around the island who were fishing in different areas at different seasons in succession – is, if suggestive, in the end unprovable. But the middens are very large, and were clearly used over a very long period of time. Like those in Denmark and elsewhere, it is reasonable to assume that the successive users were of the same familial descent, and went to these places and temporarily settled there because those areas were known to be productive of food. They discharged their waste onto the same middens because they used the same methods of housekeeping, and these were known to be effective from millennia of experience. So a Mesolithic lifestyle could be nomadic, just as was that of the Ice Age hunters, but the constant re-use of the same site, or series of sites – as happened also in Denmark and Japan – does imply that the same community returned repeatedly to exploit the same foraging areas over a long period, perhaps over several centuries, even millennia. Despite all the alternatives and questions and puzzles, the easiest explanation for the finds is that a single community, even a single extended family, occupied the island for a very long time, living in a way which became traditional – a Mesolithic foraging family, that is. The nomadism of the Mesolithic was thus in a much more restricted range than in the preceding Palaeolithic, and the territories of particular groups were therefore more restricted also, and so presumably more valued and more exclusive; foraging over the same territory regularly was what took place.

Comparisons

The same sort of remains can be found in other Hebridean islands, in Galloway in the south-west, and on the east coast of Scotland, though the seasonal sequences are only clear at Oronsay. An obvious alternative to this cycle of occupations would be to move repeatedly to a more productive site, without returning to those already occupied. Until 9,000 or 8,000 years ago the North Sea was dry, and the people could hunt over its plains and fish its rivers, and this was no doubt a more productive environment than a mountainous Scotland which was just escaping from the ice. After the loss of Doggerland their hunting area was restricted to the British Isles (or to the European continent, if they moved in that direction). Recourse was had to such foods as mussels and limpets to eke out the food supply which now did not include access to the greater beasts of prey. The same development can be seen in Denmark, where the less extreme climate which succeeded the Ice Age, and the calmer winters, may have contributed to a more stable lifestyle. In both western Scotland and Denmark the middens also suggest a regular movement from one hunting and gathering area to another through the seasons; given the recent drowning of Doggerland, it could be that the proto-Hebridean Islanders, the new Norwegians, and the proto-Danes were all cousins; in Japan the foragers operated in possibly an even more regular, even regulated and restricted fashion, which may be one of the sources for the formulation of Japanese society even now.

These Danish fisherfolk and the seasonal inhabitants of Oronsay had actually made only the most minimal adaptation to their lifestyle in the new conditions of the end of the Ice Age. They were still hunters, still gatherers, still fishermen, though they had both managed to locate a territory through which they foraged and which was productive all the year round and which so allowed them to take up a permanent, or near-permanent, residence. They had reacted in the same way, if in different places, to the rise in the sea level. Their new life meant that they did not need to follow the herds of reindeer or other animals as they migrated. They did not, yet, discard entirely their own migratory habits, but they concentrated them into a fairly small area. It seems probable that they moved with the seasons, fishing when and where the fish were plentiful, and gathering shellfish when they were available, clubbing seals on the shore when they were at their most vulnerable, hunting land animals in other seasons, and returned to the same place on a regular cycle. The houses they built were clearly temporary, thrown up in a brief time, or even perhaps tents, and abandoned them as they moved on. The remains they left behind in the shell mounds have attracted the most attention, simply because of the size of the middens, and their age, and for the information to be extracted from them. They were still essentially hunters and gatherers of very wide range of foods, but they were now compelled to concentrate their activities on smaller prey and in a smaller area; this insistence on the most easily acquirable lands came about because they were available: they were responding to the new climate. So the one serious change in their lives which had come about was that they concentrated very largely on one type of food. This was, of course, not the case in Japan, which was a much more productive environment, or in Denmark, but all three of these groups of people had reacted in a similar manner.

They had arrived by sea, and therefore in boats, perhaps the same sort of hide covered boats which were used by their early contemporaries who went from Doggerland to Norway. The occupation of other islands at about the same time suggests that they arrived at their various new homes – islands and the mainland – as a result of an organised migration from elsewhere, again like the one I have imagined for the settlers of the Norwegian coastlands. They presumably arrived from the south, which must mean Ireland or western England or Wales (to use the modern names). But we do not know.

One further aspect of these people requires attention, and it is implied in the evidence from Japan and Denmark and Oronsay. These middens were permanent features in the landscape, at least by the time they had reached their visible size. They were large and obvious, particularly so when still in use, when their white bones and shells will have been very easy to see in the green and brown lands. It is suggested that they were repeatedly used by the same families throughout many generations, that is, that they were in a sense owned by those families. In Japan, the concentration of groups into the valleys, where the permanent occupation sites have been found, similarly implies territoriality. The proto-Danes who constructed the fish traps must have regarded the traps and their product as primarily belonging to their own group, as a product of their own work and preparation.

To some extent this attitude of the exclusive use of particular resources may be noted in the Palaeolithic period, where it seems that the caves and campsites were repeatedly used, probably by the same family, as at Meiendorf/Stellmoor where it seems obvious that the same hunting group returned repeatedly to the same summer camps over a period of at least a quarter of a century. And in the Mesolithic there appears for the first time the phenomenon of cemeteries, where a number of dead – presumably of the same hunting group and family – were buried in the same constricted space, perhaps over several generations; again this implies the 'ownership', or perhaps proprietorship, of a particular territory. The burials found in the Oronsay midden would similarly imply ownership by the family of the deceased.

This idea of territoriality, which clearly has within it the concepts of ownership and exclusivity, is a characteristic of human behaviour, but it is in the Mesolithic that it becomes fully visible; in hunter-gatherer activities the Palaeolithic hunters no doubt had some sense of a region in which they had the almost exclusive hunting rights as against other families of hunters, but these would be large areas, quite unpoliceable, though evidence of violent deaths in that period might suggest that enforcement attempts were made. It was because of the development of foraging as a technique of food production that permanent settlements in one place, and so detailed territorial possessiveness, evolved and became more widely accepted; the smaller, more restricted and definable areas they used as their resource base could be controlled, and outsiders had to be kept out so as to conserve the resources for the family or community. The restricted areas also meant a restricted resource base, and so defending the area would help to ensure food on the table. One of the consequences was the development of hostilities between groups of foragers. The Spanish cave paintings exemplify all this – the exclusive rights to honey from a beehive, the battle between two groups, armed with bows and arrows and spears, which fight would almost certainly be fatal for someone (see Gallery III).

In addition, if we can accept the idea that the same group used the same areas for foraging over several, perhaps many, generations, it is a situation in which distinct languages will develop. Even before the Mesolithic, if the Ahrensberg culture in Western Europe was that of the people who lived in the Ice Age in northern Germany and in Doggerland (not proven, of course), these will have spoken a single language. When they were scattered to Denmark, Norway, the British Isles, the original common language will have fragmented, and new languages developed in each region to describe and explain their new homelands, its resources – and its boundaries – though none of these languages survive at the present, of course, though they were presumably the ancestors of the later Scandinavian languages. It is worth recalling that the Mesolithic period lasted several thousands of years, quite long enough for such language developments to take place.

As yet, in the Mesolithic period, territoriality and ownership was still a somewhat vague concept, limited to exclusive control of a foraging area of some size, and perhaps one not yet wholly exclusive, but violent deaths of humans in that period, as earlier, indicate that a further implication of ownership also now existed – war.

So we may theorize that the Mesolithic saw the development of the idea of territoriality – of an area which was regarded as home to a particular group of people – of ownership (of land and/or resources) – of constructed and organised, and therefore owned, economic resources, such as the Danish fish traps – and of distinct languages, which would even more clearly mark out strangers as 'others'. Put all these together and the result is a series of, probably fairly small, nations spread over Europe. Much the same must have happened in Japan as well.

India

In India Mesolithic period remains have been located throughout the subcontinent (Fig. 45), but a particular concentration of such sites has been investigated in the north, in the Vindhya Hills and the Ganges plain. All of these are in a relatively small area not far from Allahabad (Fig. 46). The use of small microliths as tools had already developed in the last period of the Ice Age, as it had in Palaeolithic Europe and eastern Asia and elsewhere. The Mesolithic is marked by the successors of the wide-ranging hunters of the Palaeolithic, who lived in semi-permanent habitations. That is, much the same development took place in India as in Japan and Western Europe as a result of the changing climate, which in India saw a wetter regime replacing an arid period, and so the growth of resources took place, and this made it unnecessary to range so widely to gain food. Again, we can see the same reaction by widely separated groups to the same problem and to the same new opportunities.

The relevant sites in the Gangetic plain range in date from 12,000 to 6,000 years ago. The earliest, Sarai Nahal Rai, about fifty kilometres north of Allahabad in the open plain, was occupied at the time when the water run-off from the Himalayas was particularly strong, that is, at the time when the Tibetan ice cap was being drastically reduced by the warmer temperatures. Two other sites, Mahadaha and Damdama, are a few kilometres away to the north-east, but still in the plain; these are rather later, and were occupied when the heavy run-off of water from the Himalayas had been reduced so that some of the former rivers had dwindled to streams, and had left abandoned river meanders as oxbow lakes. All three sites show that the people hunted for their living, their prey being wild cattle, goats, sheep, and deer, but also rhinoceros, pigs, turtles, and bison; fish was an occasional variation. They butchered their animals close to their dwellings, and at Mahadaha they threw the unwanted remains into the nearby lake. There was little evidence of vegetable foods being consumed, except for parched wild grains, but they were no doubt eaten as available. It would seem that the product of the hunting was plentiful enough – the wide range of animals, and the size of many of them, suggests this – that the people could settle down for an extended stay in all these places.

Their skeletons show that they were tall and robust, with large heads and powerful jaws, well adapted to chewing tough meat. An elaborate series of measurements of the 41 skeletons excavated at Damdama have shown that, as might be expected, most of them died fairly young. Half the dead lived for at least 35 years, but only two were over 50 at death; this also meant that half died before reaching the age of 35, and in fact 16 out of the 39 did not reach the age of 25 years. These younger deaths are heavily biased towards young men, and the only two to reach over 50 years of age were women. The absence of children under teenage years in the cemetery (except for two whose sex could not be identified) suggests that they were buried elsewhere, if at all. This distribution of ages at death may be taken as typical of any Mesolithic population, and perhaps of any preceding Palaeolithic group as well.

Their food was caught mainly by throwing spears and stones, judging by the development of their arm and shoulder muscles. In many ways this concentration on hunting large animals shows that they were still living

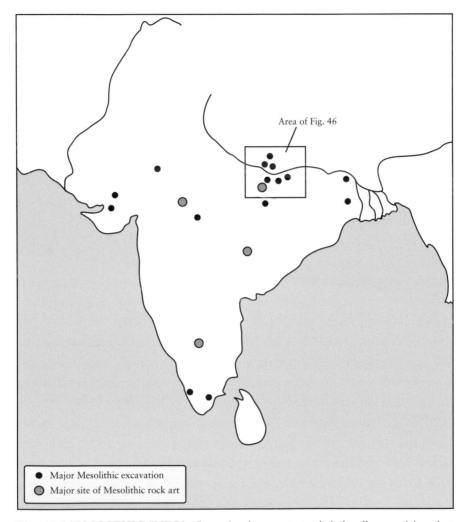

Fig. 45. MESOLITHIC INDIA. In such a huge country it is hardly surprising that archaeological work has not been intensive. The distribution of Mesolithic sites is curious, therefore, and probably reflects local interest and work. The distribution of Mesolithic rock art depends, of course, on survival and on the existence of the caves and rock shelters which were the favoured places of the artists.

a Palaeolithic lifestyle, except for the use of microliths – and, crucially, their sedentary nature.

There is evidence in all three sites of clay being used in hearths and in plastered floors, but there is no pottery. Elsewhere, at another site in the Vindhya Hills, Chopani Mando, a site which produced evidence

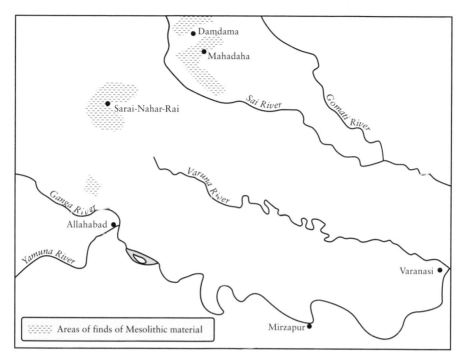

Fig. 46. MESOLITHIC SITES IN THE GANGETIC PLAIN. The three sites discussed in the text are all in this valley, where the rivers were very powerful at the time of the melting, but which had calmed down somewhat later. (They still flood annually, of course).

that it had been used intermittently for about 10,000 years, and so since well back in the Palaeolithic, there were traces of huts from the 'early Mesolithic' period. In all these sites graves were found which showed that the dead had been buried with some care. In a few instances the graves were lined with soft earth before the bodies were interred and in others, especially at Sarai Nahar Rai, the dead were sent on their way with gifts of microliths made of especially attractive materials such as carnelian, or ivory. Here is evidence again of their attachment to a particular place; mobile nomadic hunters are unlikely to maintain cemeteries; their dead are more likely to be simply discarded, or buried singly. The care with which these Mesolithic people were buried at all the sites, and that there were several of them in cemeteries (and no doubt there were more which have not been discovered) is the result of the long occupation of the site by a particular community, and its exploitation of the region – in fact, ownership. These are all traits which indicate that these hunters' villages were all of one culture.

India is a country which is particularly rich in the rock art of the Mesolithic period. It appears in many parts, but especially in the upper valley of the Narbada River in central India, some way south of the homes of the hunters of the Ganges valley. One reason for the number of these paintings is presumably that the Mesolithic lasted longer in India than in most other areas – in the eastern and central areas until about 5,000 years ago – before being overtaken by agriculture. The same argument may be made for Japan, with its development of the ceramic arts. As a result of the cave paintings – or perhaps drawings might be the best description – we have a unique record of the life of Mesolithic period in India. Many of the more general assumptions behind these pictures may be transferred to other Mesolithic groups, because they are the normal activities of human beings, but the illustrations are useful in reminding us of this (see Gallery V).

They were hunters, of course, and they are shown tackling even the largest animals, such as rhinoceros, and the range of animals hunted is confirmed by the archaeological findings. They operated singly and at times in co-ordination. They chased and they trapped, at times preparing large fenced areas for the reception of their prey – just as the proto-Danes build fishing traps – and drove their prey into a limited area for ease of capturing. They lived in houses of a fairly substantial construction, which could perhaps last several years, and they lived in nuclear family groups of parents and children, one of which is depicted having a family meal together, the parents facing each other across a table, the children playing on the floor. The food seems to have been prepared in the homes by the women – no surprise there – and some of the pictures give us a good idea of the range of equipment they could use. They were musical and reinforced the communal spirit by dancing. They wore masks, sometimes in the dance, but also in the hunt, perhaps as a disguise, perhaps for more spiritual reasons – as did the Star Carr people, and others – another legacy perhaps from the Palaeolithic hunters.

These pictures, and the use of substantial houses and of cemeteries, all imply a sedentary life, and this means that, even if they were big-game hunters as in the Palaeolithic, the wealth of the resources around them permitted them to live in permanent homes and villages. Their life was, that is, a version of foraging in which the emphasis lay in hunting the resources of a particular area closely around them, which was probably fairly limited, and operating from a permanent base. Their life may have been one of hunting, but they lived in a very similar way to the Oronsay people and the proto-Danes, who both emphasised the products of the sea, and the Jomon Japanese, who mixed all sources.

Peru

A completely different environment was tackled by groups of people who lived on the coast of southern Peru in South America about the same time as the proto-Indians lived in the Gangetic plain and the proto-Danes fished in Danish waters. Two sites in particular shed light on their lives. Quebrada Tacahuay and Quebrada Jaguay were places occupied between 12,500 and 10,500 years ago (and so before the decisive opening of the ice-free corridor between Alaska/Beringia and the rest of the continent). These places are now small riverine plains looking out onto the Pacific Ocean, though at the time of their occupation the sea level was lower so the ocean was a good deal further away than it is now; it is likely therefore that there were other settlements closer to the coast at the time, and the resources were considerably greater in fruits and root crops than at present. It may also be assumed that other, similar, valleys to north and south housed the same sorts of communities. The inland background was the steep mountains rising to the Andes (Fig. 47). They were thus in a very similar situation to their contemporaries in the Japan of the Jomon foraging period, in a limited area and facing the sea, though their foraging area was even more restricted by the mountains and the steep valley sides, and, like the Indian sites, they were in river valleys which were subject to a strong run-off of meltwater from the frozen mountains.

The people of these Peruvian valleys lived mainly by fishing from the shore and by gathering shellfish – a combination of Oronsay and Denmark, in effect. They presumably hunted what they could, but few or no remains provide any information about that – the steep land is not conducive to hunting or to the presence of large animals, and Peru did not have any of the latter. The plain between their villages and the sea

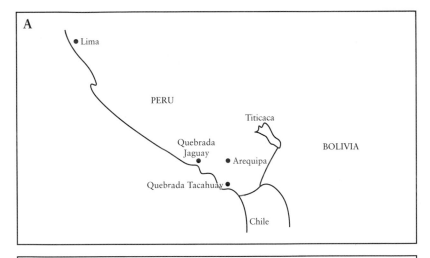

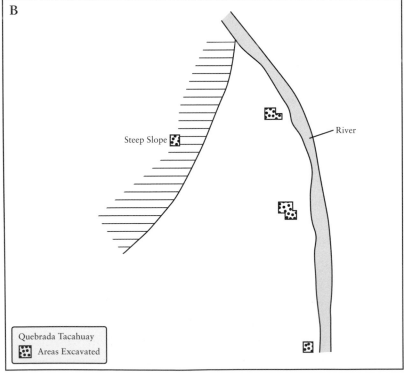

Fig. 47. TWO PERUVIAN EARLY SITES. Quebrada Jaguay and Quebrada Tacahuay were small settlements of fishermen and foragers. The settlements were close to the coast where streams came out of the mountains (A) ('Quebrada' means river). The plan of the site at Tacahuay (B) shows the situation – the sea was a few kilometres away. The settlement was eventually destroyed by a flood, probably caused by the extra storminess or an 'el-Niño event'. For a time, however, the inhabitants adopted a sedentary lifestyle.

was a source for vegetable foods (which rarely leave much in the way of remains for archaeologists to discover). At Quebrada Jaguay the main prey was a fish called drums, along with molluscs; at Quebrada Tacahuay the main food was anchovies and seabirds, especially cormorants. Fishing was done, it seems, by nets made from twine fashioned from the fur of such animals as llamas, and was done possibly directly from the shore, for some knotted cords were found, though there was nothing in the way of boats; they could well have used floating logs, however (Peru is the home of the balsa tree). At Quebrada Tacahuay the earliest occupation was ended by a flood from inland which covered the site with debris; at Quebrada Jaguay occupation seems simply to have ended.

Other more or less similar maritime gathering societies have been located along other coasts of the Americas – in California, in Tierra del Fuego, and in the Maritimes of Canada, mainly gathering shellfish along the shore, all very similar to the villages on Oronsay. In the interior of North America, by contrast, this was the time of the hunters of the great animals, in archaeological terms the producers and users of the 'Clovis' and 'Folsom' artefacts. These were nomadic, whereas those who chose to exploit the seas were necessarily largely sedentary.

Foraging as a Solution

All these societies – in Denmark and Scotland, in the Gangetic plain of India, on the Pacific coast of Peru, in Japan – were new, above all in the sense that they found it possible to settle in one place for lengthy periods of time in order to exploit the surrounding areas, sea or land, so that they could acquire the resources to live well. Since they maintained these lifestyles for millennia in all cases, we may assume that they felt that it was a life they were comfortable with, and saw no reason to change. In the process they had all to develop skills in hut building and in building more permanent homes for themselves, in making hearths for fires, using clay (for hearths or later for pottery in some cases), and in manufacturing tools and weapons from small pieces of stone. In many cases they continued the widespread artistic tradition already evident in the Ice Age: the numerous rock art paintings and drawings in India are examples, as is the early Japanese pottery, which, from the very earliest examples, was already being decorated with intricate and eye-pleasing designs. Mesolithic Europeans also produced rock paintings and carvings in stone and wood.

There was also the widespread practice of honouring the dead with careful burial, something which, given that these communities were now sedentary was necessary for hygienic reasons if for no other. But such a practice permitted displays of affection and so placing parting gifts in the graves or on the bodies – which had not been uncommon earlier, of course, as the careful burial of the Lake Mungo man demonstrates. The whole, however, also implies a spiritual dimension to the lives of all these people, or at least a belief in the continuation of life after death, and, because the grave's location was known and present, a collective memory of particularly notable dead people, and family members. This was, no doubt, based in part on superstitions such as guesses as to the source of thunder and lightning, or floods, guesses which, by attributing such events to gods, were clearly mistaken. Yet the burial of dead husbands and wives and comrades and parents with care and with gifts, and, in Japan, the burial of domesticated dogs, argues not only a set of beliefs which may be defined as religious, but also, perhaps more directly and more powerfully, a strong feeling of a continuing society. Careful honouring of the dead predecessors of the society linked the past with the present and looked on to the future, when the present members would also be honoured and remembered. This development is not really new, nor was it confined to the sedentary people, but it was surely an attitude which was much fostered by each group's continued occupation of a fairly small area or a particular place; in the same way, this will also have enhanced the feeling of 'home' and the 'ownership' of a particular territory.

The change which all these groups underwent was a collective response to the effects of the change in the climate. This is implied by the fact that the same reactions took place at more or less the same stage in the climate change by people in Japan, on the north-west coast of North America, in western South America, on the Gangetic plain in India, on the islands and coasts of Denmark, and in Hebridean Scotland – all these developments were almost contemporaneous. (The alternative would be to suggest that there had been communication between these groups, with these particular ideas being transmitted across continents and oceans, which seems highly unlikely.)

The Mesolithic is primarily identified by archaeologists with the production of small bladed tools – the 'microblades' of Japan and the North-west American coast, the 'microliths' elsewhere – but it was a time of much greater significance than that, for it was also marked by the settlement in one place for more or less lengthy periods of time of many groups of people who lived by foraging in a restricted territory. This was a development which took place not just in Japan and the North-west American coast, where it is perhaps most obvious and longest lived, but in Europe, India, and South America as well. Again, unless we are to assume widespread communication and discussion amongst these groups of people on the topic, we have to assume that this took place more or less spontaneously among a wide variety of peoples who were acting in response, with care and with gifts, to the same problems of the global changes in climate which were taking place in all these regions.

Further Reading

The early history of pottery in eastern Asia is summarised by Irina S. Zhushchikovskaya, *Prehistoric Pottery Making of the Russian Far East*, translated and edited by Richard L. Bland and C. Melvin Aikens, British Archaeological Reports S1434, Oxford 2005, especially in the summaries in Chapters 1 and 5.

For an accessible account of Japanese Jomon archaeology see G.N. Barnes, *China, Korea and Japan, the Rise of Civilisation in East Asia*, London 1993, Chapter 5; for more detail consult C. Melvin Aikens and Takayasu Higuchi, *The Prehistory of Japan*, New York 1982, and Keiji Imamura, *Prehistoric Japan, New Perspectives on Insular East Asia*, London 1996.

For the Oronsay excavation see P.A. Mellars, 'Excavation and Economic Analysis of Mesolithic Shell Middens on the Island of Oronsay (Hebrides)', *Scottish Archaeological Forum* 9, 1977, supplemented by an article by Mellars and M.P. Richards in *Antiquity* 72, 1998, interpreting the 'Stable Isotopes and the Seasonality of the Oronsay Middens'; and, for a wider view, Bill Finlayson, 'Understanding the Initial Colonisation of Scotland', *Antiquity* 73, 1999, 879–884. The Scottish evidence generally is detailed in Graeme Warren, *Mesolithic Lives in Scotland*, Stroud 2004.

The Indian evidence is well summarised by D.K. Chakrabarti in the *Oxford Companion to Indian Archaeology*, New Delhi 2005, Chapter 5; see also Walter A. Fairservis jr, *The Roots of Ancient India*, 2nd ed., Chicago 1975, and D.P. Agrawal, *The Archaeology of India*, Scandinavian Institute of Asian Studies, monograph series no 46, London and Malmo 1982. The cemetery evidence from Damdama is extensively detailed in John R. Lukacs, Jagannath Pal *et al.*, *Holocene Foragers of North India, the Bioarchaeology of Mesolithic Damdama*, British Archaeological Reports S2783, Oxford 2016. The rock art is reproduced in a delightful book, Erwin Neumayer, *Prehistoric Indian Rock Paintings*, Delhi 1983.

The South American cases are published, briefly, in *Science* 181, 1989, in summary articles by D.K. Keefer and D.M. Sandweiss and others; a wider perspective on coastal settlement is provided by articles in *American Antiquity* in 1989 by A.L. Martinez on sites in Chile, and in 2001 by D.C. Rich on California.

Gallery V

Mesolithic Life in India

The rock art of India is perhaps the most eloquent of any in terms of the life of the people in the millennia following the Ice Age. The Mesolithic in India was almost as long as in Japan, for agriculture did not become adopted until 5,000 years ago, and then only in the Indus Valley at first. Until then the people in that enormous land had existed as hunters and gatherers, though, like the foragers, they tended to be sedentary. The art they painted and inscribed on the walls of their caves and rock shelters shows not only their hunting and gathering life, but also their times of relaxation and, most unusually, aspects of their home life.

Hunting is depicted in several ways. An early painting from Lakhajoar, near Bhopal (V.1), shows a man wearing a loin cloth (blowing out behind him), and a feather headdress, about to spear a fleeing cow. The curious drawing of the heads (identical for man and beast) is an indication that the painting is very early. The man carries several spare spears and wears feather armlets at his elbows. Other animals depicted, equally early but more accurately, include cattle and a horse (V.2) and elsewhere elephants, boars, and so on.

In a group of three archers (V.3), one is depicted larger than the others, and has a much larger bow; his headdress may indicate his greater authority. Again, the greater rudimentary heads suggest an early version.

They were skilled hunters, proud of their skills, as the attitude of the men in V.3 suggests, and they cooperated and organised their hunts to minimise effort. V.4 shows three men spearing small cows, perhaps calves; the two larger animals may not be part of the composition. They also, like the Danish fishermen (and elsewhere), used artificial obstacles to trap their prey, such as the deer in V.5 – again, the hunters' heads show this is an early painting.

Meat was only part of the diet, and fruits and vegetables were also gathered. In V.6, from Bhimbetka, three people are gathering fruit from a tree, one of them climbing. In V.7, the fruits of the expedition are being carried home in a bag or basket; these are actually painted on the same wall, only 24 centimetres apart, and might represent a single event.

The last example (V.9) is virtually unique in all the cave paintings in showing a family at home at dinner. The woman, apparently pregnant, rests as the man appears to be preparing the meal; their child plays. All are enclosed in a hut made of a wooden framework, thatched.

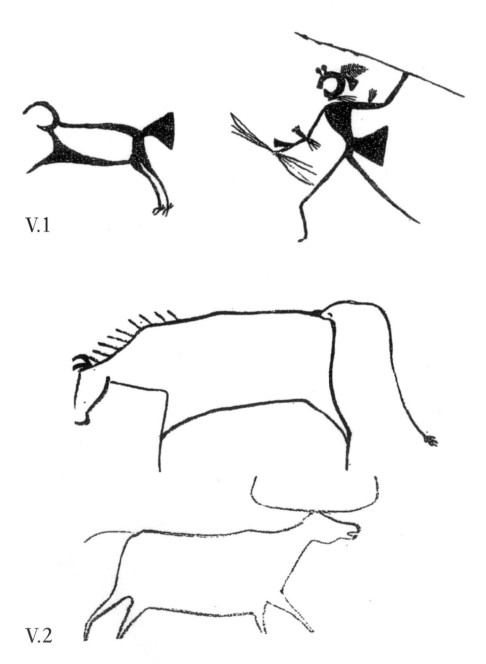

V.1

V.2

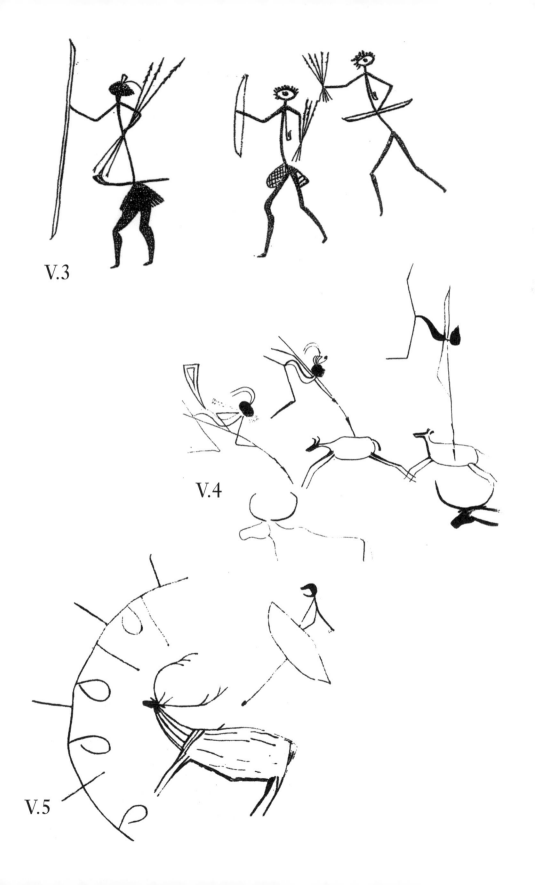

V.3

V.4

V.5

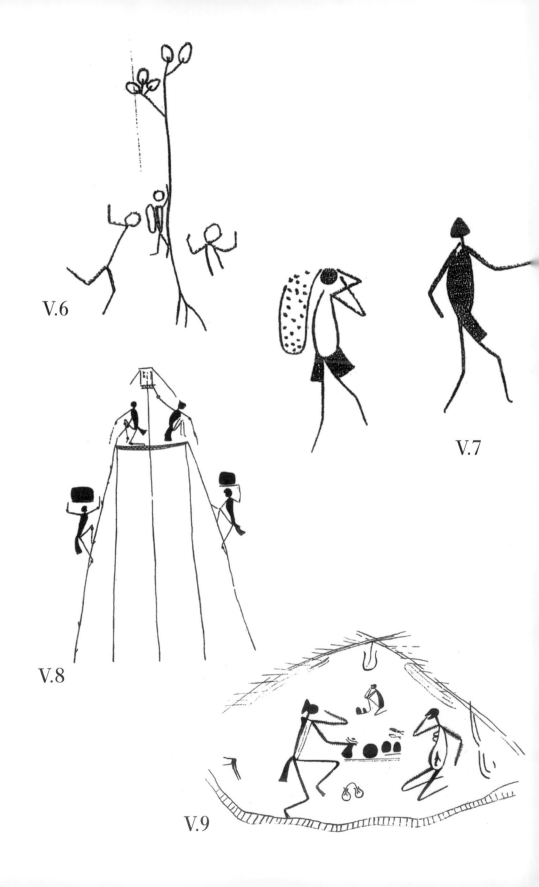

V.6

V.7

V.8

V.9

Chapter 6

Drought

During the Ice Age India had been an arid land, the monsoon diverted, vivifying moisture held in the ice, and the sea level lower. As the climate warmed so the land gradually developed the climate it has now, with the usual fluctuations resulting from the erratic ending of the Ice Age. The foragers in the Gangetic plain – and elsewhere in the subcontinent – lived under a much wetter climate regime than their Palaeolithic ancestors, though the monsoon system – which had been diverted south during the Ice Age, and had been weakened – rather skews rainfall to a small part of the year. The sub-continent was, indeed, probably wetter than at present for several millennia (as in the Gulf), and the rivers were even more powerful than now, being fed by the melting Himalayan and Tibetan ice.

Australia: the Changes

Elsewhere in the world the Ice Age regime of drought continued, though somewhat modified in extent and intensity. In Australia the desert centre of the continent had expanded during the later millennia of the Ice Age (Fig. 12). Of course, the joint Australia-plus-New Guinea-plus-Tasmania island-continent of 'Sahul' was substantially larger than the separate islands are now because of the lower sea levels and the consequent exposure of shallow sea areas, but the desert part of Australia was then spread much more widely even than today. The ice caps in the continent were relatively small, one in Tasmania, and a fairly small one on the mountains in New South Wales. Curiously there was more ice on the New Guinea Mountains, which are twice the height of the Australian Alps, than on either of these. By this time the inhabitants of the great continental island had partially adapted to the dry conditions, in that they had located those places where the water was normally available, and were able to hunt throughout the surrounding lands.

They had also had to adapt to the disappearance of substantial numbers of the animals which had been in the land when they had arrived: this was another 'great extinction' such as that which happened at the end of the Last Glacial Maximum in North America. At least forty species died out, principally the larger animals. The process took a long time, which is also to say that 'extinction' is a process, an ongoing business spread over millennia, and so any single cause is highly unlikely. As ever, humans have been blamed, though this largely is a product of modern guilt, and the assumption that men have the power to do so. The basic cause was probably a natural process by which individual species simply died out, but the effects can be accelerated, especially given the fact that it was the larger animals which died out, by such means as the drying out of the continent, consequent on the change in climate. That is, both the human beings and the animals were faced with the same problem, just as in Europe while the Ice Age ended. In Australia the animals failed to adapt to the expanded desert, but the more flexible human animals were able to cope with the new conditions. Animals requiring copious quantities of water failed: men no doubt helped the process along – in such circumstances hunters could simply ambush their prey at water holes – but this could only be a contributory factor, if that. But by the time the Ice Age ended, the greater beasts had gone. As a result, of course, life for the Australian hunters became that much harder.

The increase in the world temperature generally and the melting of the ice caused enormous changes, as great as those in any other continent. The sea level, of course, rose, so that wide areas of low lands around the coasts of the southern continent vanished. This was particularly serious in the north, where the connection between Australia and New Guinea was progressively reduced from about 12,000 years ago, and was finally severed about 4,000 years later; the connection between Australia and Tasmania was broken at much the same time (see Chapter 4). The newly flooded lands in the north were a huge area, which is now occupied by the Arafura Sea and the Gulf of Carpentaria (Fig. 3). (The area which was flooded in Sundaland, further to the north, was even greater, in effect replacing a continent with an archipelago.) The inundated land had been almost flat so the flooding, when the sea reached the right level, will have taken place relatively quickly. One suggestion is that the sea

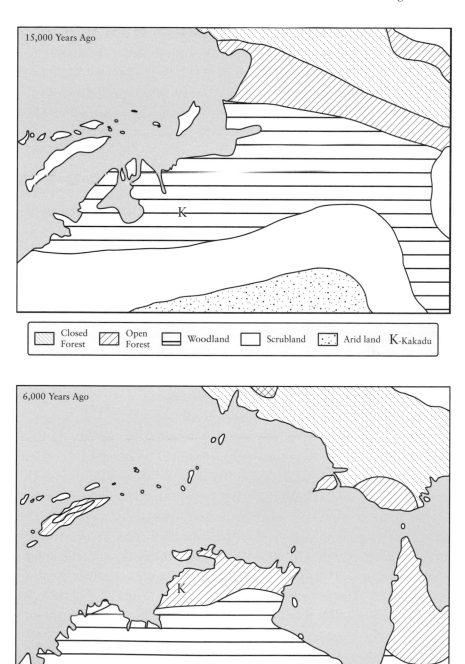

Fig. 48. CHANGES IN THE AUSTRALIAN NORTH. The north of Australia changed dramatically with the flooding of the Sahul shelf. In New Guinea the great forest extended over all the island (except the mountainous interior). In Australia the increased rainfall pushed the desert southwards. Overall, however, the woodland decreased in size with the loss of the lowlands.

at times advanced four metres in a year; another is that the shoreline advanced (or retreated) by up 100 kilometres in a generation, though such generalisations omit the small variations in height within the flooded land – the floods will have been much less predictable there – and the variation in the rate of sea level rise. Nevertheless, these advances were speedy enough that humans and animals will have been caught by surprise, especially since the advance will have been erratic, failing at times, and at others very fast. There can be no doubt that plenty of people died, particularly along the river valleys and in the lowest lands.

The three islands developed contrasting climates. Tasmania, cool and wet, was and remained heavily forested, with the inhabitants, who were isolated from Australia now by the flooding of the Bass Strait (Fig. 37), organising their lives as successful hunters and gatherers of the classic type. New Guinea was (and is) tropical, with plenty of rain. Its internal forests remained dense and superficially inhospitable, but they were fruitful in foods. (I shall return to this region and subject in the next chapter.) As a result of the climatic changes Australia became essentially a largely desert continent with a rim of less dry lands, even wet enough to be tropical along the northern and eastern coast, and wet enough along parts of the south and west coasts – as the climate is now. The dampness and size of these areas varied, from tropical heat and damp in the north to a close-to-desert dryness in the south and west. The desert land reached long stretches of the coast on the north-west and the south, and yet the far north of Queensland and the Northern Territory was similar to New Guinea in many ways and was rich in both vegetation and water. The vegetation on the east coast changes as one moves south, from a tropical lushness in the north to a dry-ish Mediterranean type in the south. People lived in and prospered in all these lands, from the wet north to the dry centre. This is a powerful mark of the adaptability of humans, as much as the successful hunters originally from Africa who moved to and lived in Ice Age Europe.

The rising of the sea level was accompanied both by an encroachment of the sea on to the land margins, but also by a certain reduction of the area of the desert land, as the increase in the moisture in the atmosphere and the penetration of moist ocean air inland brought moisture to some of the lands which had been desert at the end of the Last Glacial Maximum.

The proportion of desert land to the whole did not really change much, however. The changes in Australia were thus not all that radical: the flooding of the lowlands was, of course, serious, though, since these are now under water and have scarcely been explored archaeologically, it is impossible to say just how big a disaster the loss of this land was to the inhabitants.

Arnhem land – the Kakadu

The greatest change came in the north, where a huge area was converted from land to sea. Much of the flooded land had, it seems, been wooded in various ways, ranging from shrubland to tropical forest, during the Last Glacial Maximum. The effect of the change was to increase the quantity of moisture reaching the northern areas of Australia – Arnhem Land and northern Queensland – so that the successive bands of woodland expanded southwards – the sequence is open forest, woodland, shrubland (or savannah), and finally desert. So this expansion of woodland and so on occurred even as much other woodland was being lost to the sea. The initial increase in precipitation (which was seen also in India and North America and the Gulf, and no doubt elsewhere) was not maintained, so that by about 6,000 years ago the present vegetation regime in the north had become stabilised (Fig. 48).

In the Kakadu region of Arnhem Land, a region which has been well studied, it seems that the encroaching sea initially forced the human population to occupy the land more densely and so to exploit it more intensively (Fig. 49). The people had a wider, or better, set of food resources when the valleys became partially flooded by the rising sea, an indication that these changes were not always detrimental. But the process of adjustment was painful and at times violent, and the increased rainfall produced annual floods which also had to be coped with.

The intensification of the human occupation is indicated by the larger number of sites which became occupied as the change continued, from just one or two rock shelters which were in use about 15,000 years ago (during the Last Glacial Maximum) to the six or seven about 8,000 years ago (at the end of the cold period of the Younger Dryas). These later shelters were also more intensively occupied, with deeper sediments and

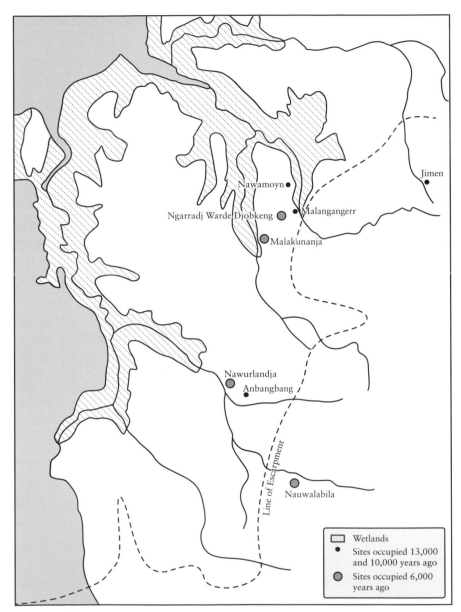

Fig. 49. SITES IN THE KAKADU. The flooding of Sahul also produced a new environment in the north, where the area now called Kakadu became close to the new coast. Wetlands came into existence along the rivers, and these increased the resources available to the local inhabitants, and as a result of this, and of the loss of the Sahul land the number of occupied sites – caves and rock shelters mainly – increased from three to eight. The older ones were also more intensively used by larger numbers of people. One result was a competition for resources which seems to have involved fighting.

more numerous stone tools being discarded, along with the bones of the foods they consumed. That is, instead of intermittent visits, more and more families were inhabiting the shelters and were remaining for longer periods in them. The population would seem to have increased from about two families to a dozen or more. One result was a new exploitation of shore-side resources, shellfish and fish easily caught from the shore, and this wider diet is signposted by the beginning of the build-up of shell middens at the new shores. (It was, it will be seen, a sort of Sedentary Foraging regime which had developed.)

The difficulty of the adjustment is revealed in the art of the time, preserved in the shelters and caves. The region has some of the earliest evidence for cave painting on the continent. The relatively few inhabitants in the pre-flood period mainly drew pictures of animals, often almost life-size – they are, of course, examples of their hunting prey. But the increasing moisture drove the desert edge back, and with the changes the traditionally hunted large and largish animals disappeared to other areas. The growth of the forest therefore forced a change in hunting methods, so that, for example, the boomerang, a distance weapon, became little used, for the forest was the wrong environment for it.

The increase in the population brought a new competitiveness for the available resources, and some of the rock paintings show that men were involved in fighting each other. There are scenes of men being chased by others wielding weapons, and other pictures are of men dodging thrown spears or boomerangs. This violence was probably new, for the earlier population was very thinly spread. It can best be explained by the combination of increased pressure on resources and the denser and more numerous population. (It is echoed, of course, by similar evidence from elsewhere, notably the Spanish Mesolithic paintings (Gallery III)).

The effect of the rising sea level here was therefore to deprive the people of the area of much of their hunting grounds, to crowd them – relatively speaking – into a smaller area, and to promote conflict between them, at least until a new balance of men and resources was established. These were hunters who were accustomed to the well-vegetated area in the north, and not far to the south the desert was unavailable to them; their options had narrowed, but they had not yet been compelled to change their lives very seriously.

Australia – some Islands

The north-west coast of Western Australia, to the west of Arnhem Land, gives a clear indication of the difficulties faced by the inhabitants. This is, and probably always was, a desert region. There have been a fair number of excavations in the area, most of them fairly near the modern coast, but also in the offshore islands, particularly the Monte Bello Islands (which were part of the mainland before the rise in the sea level). In the area as a whole there is a consistent gap in dated occupation during the Last Glacial Maximum, from about 17,000 to about 12,000 years ago; evidently during that time the land was too dry for even the skilled hunters of the pre-dry period to make a living from it.

The rise of the sea level took place over several millennia after about 12,000 years ago, and the concomitant increase in moisture brought animals and people back, though the area was never seriously moist, and counts as a desert today. Some of the rock shelters which had been formerly used before the intense dry of the Last Glacial Maximum were again occupied, but only on a very intermittent and occasional basis. Off the coast, the Monte Bello Islands, which had been part of a large peninsula attached to the mainland in the dry period when the sea level was lower, became, as the sea rose, a single large island first of all, and then a separate archipelago of smaller islands. The inhabitants ate fish and shellfish as a staple part of their diet – as in Oronsay and Peru and other places – but as the area of land decreased the islands were abandoned. This had happened by about 7,500 years ago. The people had either died out, or had sailed away (as in the Bass Strait Islands, rather earlier); they had, it may be noted, the ability to do this, and so had the use of boats.

The desert mainland, to which any refugees leaving the islands will have gone, was hardly an attractive alternative for people who had become used to the relative abundance of the sea. They may have moved on further – or, again, they may have died out. Wherever they went, it does not seem that they would have gone back to live in the desert, though some groups were able to make a living in the region by constantly moving and exploiting very large hunting regions. The desert, in this area at least, was winning.

In both of these regions, the Kakadu of Arnhem land and the Western Australian coast, and also along other coasts of Australia, there are clear signs, usually in the form of shell middens and the remains of living sites, where these have been excavated, that the resources of the shoreline were already being exploited from the time the sea level stabilised at its present situation. (It is therefore probable that earlier middens have been flooded and living sites been left behind when the sea level rose, and could perhaps be discovered by underwater archaeologists.) This was still a nomadic lifestyle, and there does not seem to have been a systematic concentration on exploiting the sea such as that of the proto-Danes or the Fish-eaters of the Gulf; nor is there any sign of any development of a sedentary life, except perhaps in the Kakadu for a time. On parts of the coast the shore was certainly used as a full resource, where shellfish could be harvested, though no permanent occupation was ever apparently developed. The Australians did not organise themselves as sedentary foragers, as did the Japanese, no doubt because of the greater size of their land, and the more difficulty they had in the environment they inhabited; perhaps they also preferred the more varied diet than that produced by shellfish, but in comparison with the Japanese situation they did not usually have a concentration of resources available in a relatively small area and all through the year, so constant migration was necessary. The existence of the middens, however, does imply that some communities, or families, adopted a sedentary life, or perhaps circulated through a regular series of living sites, as at Oronsay, but probably, given the desert land, on a much wider and more extensive scale.

The new shore, of course, only became the shore after several millennia of encroaching seawater. The sea has always been a resource of food, and it is probable that fishing and the collection of shellfish had been practised since the Australians reached their new home. They had come by sea, after all, and the earliest inhabitants had landed somewhere along the north-western coast of Western Australia, which is, and was, particularly dry. As the waters advanced, though, anyone who lived along the coast would need to retreat. Some people, as in the Bass Strait Islands (Chapter 4), or possibly the Monte Bello Islands, became trapped, or at least isolated, as the islands were created. No doubt in other areas, besides Arnhem Land, the situation produced a dangerous pressure on resources, and therefore conflict.

On the other hand, there is a different result: off the Queensland coast two islands came into existence about 8,000 years ago when the surrounding land was flooded – as in the Bass Strait and with the Monte Bello Islands off Western Australia. In these cases, too, people were cut off, but they survived and did not attempt to leave. They made their homes in rock shelters on the islands, which are now Hook Island and Border Island in the Whitsundays group, both of them now behind the Great Barrier Reef (which did not exist at the time of the flooding). After some 2,000 or 3,000 years the inhabitants had developed a wholly maritime lifestyle, with specially adapted weapons with which they hunted large sea mammals. The inhabitants of another island, Keppel Island, also off the Queensland coast, were similarly isolated, and this isolation persisted for so long that they developed their own dialect, which could not be understood by their mainland neighbours. These examples show that even relatively small populations can survive isolation for long periods of time if they had the technological and food resources necessary. The Bass Strait islanders clearly did not have such technology. (See Fig. 4 for these islands).

Inland there was a similar abandonment of newly-inhospitable territory, but also there was the continued occupation of land already in use. The interior lakes along Willandra Creek and nearby in western New South Wales (one of which was Lake Mungo) dried up; Lake Eyre in the south centre of the continent gradually shrank to its present size. These changes compelled some of the people who had lived and hunted in the areas to move elsewhere. They were, however, already nomadic, so the disruption to their lives was gradual and effectively under their own control; this was a similar set of responses to those of the Paleolithic hunters of Europe at the end of the Last Glacial Maximum.

The spectacular changes to the Australian coastlines to north and south, and even in parts elsewhere, were therefore not replicated in such a drastic change in the interior. The people who were affected did not react in such a decisive way as any of the other groups noted already in this book. Those along the coastline – the new coastline – did harvest the local seas, both the shellfish and the fish, but they continued their roaming nomadic life on the land, whereas the Peruvians had settled down to a sedentary life for a considerable time when placed in a similar situation. Those in the desert perforce continued their nomadic life and were able to move on when the lakes they depended on dried up, unlike the African desert dwellers to be considered next. More decisively, in the north, the contemporary developments in New Guinea (to be noted in the next chapter) had no effect, even though at the time the two islands were still joined. In Australia it would seem that the challenge of the changes was never so tough or challenging or threatening enough to force the invention of a new way of life. That is, like all the other groups we are looking at, the Australians were thoroughly conservative, and did not change simply for the sake of changing, unless forced into it; the size of their land allowed them simply to move elsewhere if the changes threatened to overwhelm them.

Sahara

The other great desert of the world, the African Sahara, shows a certain similarity with the Australian desert history in that it has increasingly desiccated since the end of the Last Glacial Maximum, but it also shows some decisive differences, both in its climate history and in the reaction of its inhabitants. During the Last Glacial Maximum, the Sahara was dry, so much so that in this period the Sahara is described as 'hyper-arid', and no human life and very few animals or vegetation existed within it. One reason for this was the generally reduced level of free moisture in the global atmosphere as a whole, frozen out by the low temperatures, a condition which only began to change with a softening of the climate which began about 14,000 years ago. This increased the precipitation in both the tropical regions and in those north of the desert areas, so that the land with the Mediterranean-type climate expanded southwards while the savannah country of grassland with occasional trees (territory also called the Sahel) along the southern boundary of the desert expanded northwards. The increase in the rainfall continued for two millennia, and then for almost 5,000 years more the Sahara was hospitable to all sorts of life. This was the period referred to by the historical meteorologists as the 'Great Humid'.

This is most spectacularly indicated by the history of a series of lakes along the southern region of the present desert. In East Africa there are

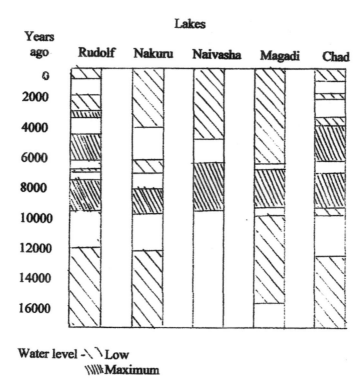

Fig. 50. THE CLIMATE CHANGES IN THE SAHARA AND EAST AFRICA.
Measurements of past lake levels indicate the sequence of rises and falls in several lakes
– Chad in the Sahara, the rest in East Africa. The climate is somewhat different in the
two areas, but it is clear that rainfall fluctuated: the great humid period between 10,000
and 8,700 years ago is clearly common to both regions. The Chad sequence also shows
the rollercoaster ride of the whole Saharan region.

Lakes Rudolf, Naivasha, and Nakura, amongst others, and in West Africa
Lake Chad. All of them were at their highest water levels between 10,000
and 8,000 years ago, after several thousands of years at unusually low
levels. Lake Chad was about fifty metres higher than at present at the
time, and consequently it was covered a very much larger area than now.
The level sank in all these lakes from about 8,000 years ago. In two of
them, Chad and Rudolf, detailed studies show that the lake basins then
filled again between 6,000 and 4,000 years ago, after which the level
sank once more. Lake Nakura also shows this change but not quite so
decisively. The levels of the lakes rose again in Chad and Rudolf about
3,000 years ago, then again between AD 1550 and 1800 (which in Europe

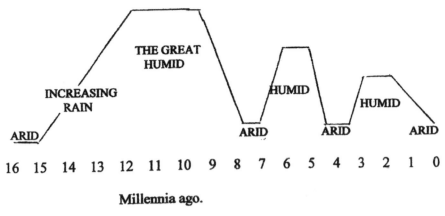

Millennia ago.

Fig. 51. SAHARAN CLIMATIC FLUCTUATIONS. In the Ice Age the Saharan region was even drier than it is now, but as global warming increased so did the rainfall, creating rivers and lakes, enlarging Lake Chad to several times its present size, and pushing the desert edge some hundreds of miles northwards. From the end of that period of warming, however, there began a series of fluctuations in which the several humid periods were decreasingly wet and the arid periods increasingly lengthy.

was the time of the 'little ice age') (Fig. 50). But each rise was smaller than the one before, and the highest levels were maintained for briefer periods in each case. Diagrammatically the climatic history of the Sahara can be shown as a wavy line, each wave less strong than the preceding one (Fig. 51).

For the animals and plants living in the Sahara, these oscillations meant alternate periods of desiccation and prosperity. In the dry periods the boundaries of the desert expanded in all directions – just as it has been doing since about AD 1800 (which, since this is one of the symptoms of the new global warming episode, shows that the change had begun in the eighteenth century, with the implication that human activity was perhaps not decisive in beginning the process – though its continuation into the present is clearly at human agency). In the wet periods, the African lakes expanded (as, in Australia, did Lake Eyre), the rivers flowed out of the mountains in what is now desert, and areas of the former desert produced forage sufficient for animal life, and animals moved in; men followed them.

Within the great expanse of the Sahara the mountainous regions attracted clouds and so were always moister than the surrounding lowlands, and in the wetter periods they often held small lakes. These areas also dried up more slowly than the lower flatter areas, and were left as damp 'islands' within the encroaching desert, while other places were left as isolated oases – 'refuges' for some of the former vegetation. The very wet time of the Great Humid period (about 12,000 to about 8,000 years ago), which followed a time of increasing rainfall from about 14,000 years ago, refreshed the water table below the desert surface, and this has remained relatively high, if slowly declining, until the present dryness. This high water table supplied the oases, and meant that they were relatively well supplied with water, while the lakes only slowly dried up. The later wetter periods halted the decline of the water table, but the intervening dry periods enforced the renewed reduction in moisture. The present water table is as low as it has been at any time in the past fourteen millennia – that is, the effects of the global warming which began with the ending of the Last Glacial Maximum have been evident in the Sahara region since then. It is a region whose very sensitivity to alterations in moisture have revealed the effect of that warming for much longer than elsewhere. Note that none of this was due to human activities.

This water regime meant that the Sahara's inhospitable nature ceased to be so extreme with the arrival of rains, though it took a substantial time, centuries perhaps, for vegetation and then animals to penetrate into the huge area of country involved. When the later dry periods struck, which probably happened fairly suddenly (that is to say, in climatic terms, of course), the oases and mountains were left as isolated places and areas where some of the animals could survive, and where men could also therefore continue to live. And this change has taken place several times since the end of the Ice Age.

The evidence for this is in the form of both archaeological discoveries, and, more importantly (and once again), of the many rock drawings which have been discovered in the mountain areas of the central Sahara. Considerable efforts have gone into attempting to date these images, with only partial and disputed success, but it is clear that they could only have been produced in times of prosperity, which is to say, only during the periods of humidity, for they show such unlikely Saharan inhabitants as elephants, hippopotami (which can scarcely survive without being able to wallow in water up to their nostrils), giraffes, cattle, and other beasts which are more usually associated with tropical Africa and which require plenty of forage to survive, and all of these animals required access constantly, on a daily basis or more often, to substantial quantities of water. People are also depicted, hunting, dancing, and fighting, in the usual ways; the paintings therefore were executed in one of the humid periods (Gallery VI).

The dating of these drawings is difficult and disputed. A whole series of elements of evidence have to be taken into account, and even then these generally produce only relative dating, not absolute. The drawings are usually incised into rock faces, scraped and/or hammered by men or women using stones. There are also paintings, made with the use of basic materials manufactured from coloured stones and vegetable dyes. For the drawings one of the dating elements is the colour of the patina which has accreted onto the incisions; style, of course, has to be taken into account; over-painting, and over-drawing, where the drawing is incised or painted so that it overlaps an earlier one, is another datum. All these elements need to be considered, but even then precise – that is, absolute – dating has not yet been possible. However, it has been possible to determine the sequence of styles, more or less.

The result, still somewhat tentative, of course, has been the recognition of a series of groups of paintings and drawings in different styles which are taken to be successive. The earliest is the 'Bubaline' drawings, from the Latin *bubulus*, pertaining to oxen and buffalo, which are some of the most easily recognisable subjects. These are large rock drawings, sometimes almost life-size, and are found especially in the Atlas Mountains in the north, in the Fezzan in the centre, and there are also paintings in this style in the Tassili Mountains to the south. (Note the parallel case in Australia, where the earliest rock drawings are also life-size.) There is also a tradition of 'Round-heads', drawings of humans with the heads enlarged and indicated by a circle but without features. Here the human body is indicated normally only in outline, a style which lasted into the next period.

The Bubaline style was succeeded by the 'Naturalistic' style, which is self-explanatory, though the images are now less than life-size, and are more detailed. The Round-head style also continued in the Tassili, while in some other areas this naturalistic style is missing altogether, presumably because inhabitation was interrupted by a dry period for some other reason. There are also later styles, but they were clearly contemporary with, and influenced by, the civilisations in the Nile Valley and the Mediterranean coast, and so are less than 5,000 years old; they may be ignored here. The subjects of all these earlier styles are much the same – the animals which were familiar to the artists and to their contemporaries, and the behaviour of the people.

Animal Domestication

There is, however, another aspect of these drawings. Some of the animals shown seem to be domesticated. This is where the dating of the pictures becomes important. One theory is that these belong to the early humid period, the Great Humid, and that the early Bubaline pictures were therefore painted between 12,000 and 8,000 years ago. The other theory is that they date to the second humid period, which has been called the Neolithic Humid, though this name rather begs the question of dating. If they are of the earlier Great Humid time, this makes the domestication of the animals an independent matter, one which took place in the Sahara region itself without reference to anywhere else – for the domestication of farm animals (as opposed to the dog) certainly did not take place elsewhere at that time. If the pictures are of the later period the domestication was probably acquired from outside, since domesticated cattle and sheep and goats already existed by about 7,000 or 6,000 years ago in the Levant and in the Nile Valley.

The hints from the rock drawings and paintings are confirmed by some archaeological investigations in part of the Sahara a little to the west of the Nile Valley. Several sites, Bir Kiseiba and Nabta in particular, have produced remains of cattle in association with human remains and tools. This material is dated to the ninth millennium before the present. Further west several sites in the middle Sahara (Enneri, Adrar Bous, Ti-n-Toha, and others) have produced similar material and are dated to about 6,000 years ago. A thousand years later, sites in the Nile Delta, in North Africa, and on the Nile in the Sudan all begin to show similar evidence; two more sites with such evidence, this time in East Africa, have been dated to between 5,000 and 4,000 years ago (Fig. 52).

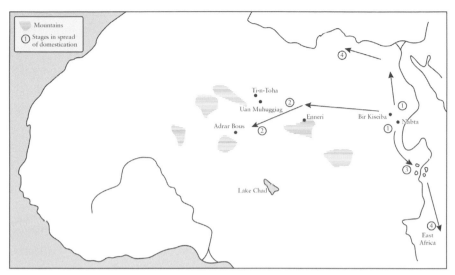

Fig. 52. CATTLE IN THE SAHARA. The early sites of cattle domestication, At Nabta and Bir Kiseiba, just west of the middle Nile, were succeeded by sites in the middle Sahara, and it is presumed that the practice of herding rather than hunting cattle extended throughout the drying savannah because of the increased stress caused by desiccation. Later the practice spread both north and south along the Nile, and before 4,000 years ago cattle were being herded in the East African grasslands.

This sequence lends itself to a clear theory: cattle were domesticated in the oases near the Nile and the practice spread through the Sahara westwards, and later southwards into East Africa and north into Egypt along the Nile Valley. The East African material is especially important, for it shows the spread of the practice of cattle herding through a 'corridor' of that land which is free of tsetse fly, and took place at the beginning of the Neolithic in the rest of sub-Saharan Africa. The theory is, however, based on too few sites and dates to be set in stone. No doubt it will be changed, and replaced by a more sophisticated model as more information and perhaps earlier dates become available. However, it does look as though a reasonable basis for the theory now exists, and the main conclusion is that the domestication of farm animals began in the Sahara.

The question of the beginning of domestication is still not wholly answered, however. The suggestion has been made that the practice may have come from Syria/Palestine into the Nile Valley across the Sinai desert (see also Chapter 7). This is a plausible theory, for other evidence indicates the existence of early connections between the two regions, but the dates of the earliest Saharan evidence – from Bir Kiseiba and Nabta – are rather too early for such a development, for they imply either a very speedy adoption of the new technique which at that time was no nearer than a thousand kilometres from Sinai, or an independent invention of domestication. It is, in this case, best to assume the latter, partly because in Syria/Palestine the animals being domesticated were the ancestors of sheep and goats, not cattle.

A theory as to why domestication occurred in the Sahara has therefore to be developed to account for it. The Great Humid period saw animals being hunted in the usual way, and the prey included wild cattle, *bos primigenius*, examples of which can be recognized in some of the paintings. As the period of high humidity faded at the end of the 'Great Humid', the animals concentrated at waterholes, at the diminishing lakes, and at oases. Men found it easy there to hunt them, a process which gradually (or perhaps suddenly) changed to herding. The bulls would no doubt be aggressive, but by killing off the most resistant animals, men would be able to reduce the herd to quiescence.

This is all a theory based on a relatively narrow range of data and sites. The archaeological material is fairly securely dated thanks to radiocarbon dating, but the correlation with the pictures is tenuous. There is, however, another theory which may assist in the problem. This is the theory that certain words in languages can indicate the society in which they were spoken, and by the use of certain specific words, present-day languages' histories can be reconstructed, a process called 'glottochronology'. It has been used with some success in the study of the colonisation of the Polynesian islands; and it has been applied, so far without much acceptance, to the issue of the arrival of the earliest people in North America – a matter which is also bedevilled by existing theories being very firmly held, even *in extremis*. In the particular case of the Sahara, it has been asserted that certain terms in the Nilo–Saharan language group are associated with domestication and its practices and that these terms existed up to 10,000 years ago. The original language is termed 'proto–northern Sudanese', and it is theorized that such terms meaning 'to milk' and 'to drive (animals)', which could only exist when domestication has taken place, were in use as early as that, and so in the

'Great Humid'. An associated language, 'proto-Sahelian', similarly had words for goats, sheep, cows, and bulls by about 6,500 years ago, but these are less specifically related to domestication, since they could also be the names for the wild animals.

If all these theories, from rock paintings, archaeology, and linguistics, can be linked, and the datings by glottochronology and by archaeology can be confirmed, we can see here the local response of the inhabitants of the Sahara to the problem of living in an environment in which the desert was encroaching relentlessly on the land: the domestication of cattle was an adaptation, therefore, to desertification caused by global warming.

It is one thing to describe a period as the 'Great Humid', but it is a very different matter actually to have lived in it. The northern half of the modern Sahara was a desert even during that time, between about 15°N and the Atlas Mountains, and it stretched right across Africa from the Nile Valley to the Atlantic Ocean. In this huge area there were, of course, some parts which were less dry than others – the mountains of the Hoggar, the Tibesti, Tassili, and others, in particular. These are the places which held on to water for longer than the lower surrounding lands, and where lakes tended to last longer, and it is in these areas that the domestic animals are depicted. The supplies of water existed, but they were limited in size and extent, and in most cases could easily be controlled by human groups. The animals dependent on these water sources could thus easily be hunted, and in the cases of cattle and sheep and goats, which are generally timid herbivores, they could be brought under control, and hence domesticated. The action was in fact an implicit bargain: the animals gained protection from their other enemies and predators, and were assured of a reliable access to food and water; the men gained a constant and ever present source of food and clothing, and, of course, they domesticated themselves as they domesticated their animals.

This chain of evidence is, of course, very fragile, but there are enough indications to propel us to the conclusion that animal domestication did take place in the Sahara region during the Great Humid period. In particular, we can theorize that it took place towards the end of that time, when the humidity was beginning to fail and the desert areas were expanding, some time, say, before about 8,000 years ago. And if so, it is a development which owed nothing to any other part of the world –

domestic animals, for example, did not reach the lower Nile Valley (that is, Egypt) until about 7,000 years ago. By then they had been in use in the Sahara lands for up to a millennium and a half – and it is presumably from the Sahara that domesticated cattle were taken into the Nile Valley; the domesticated cattle in Egypt were no doubt imported originally from the Sahara.

The oscillations of the Saharan climate obviously affected all the life of the region, and the onset of a dry period may well have killed off a substantial proportion of both the human and the animal populations (as it did in Australia), while the survivors probably withdrew to areas which were still moist or where water was available, but once it was developed, the practice of domestication of animals was a skill never lost. The desertification of the Sahel, the band of territory south of the desert which has been very obviously drying during the last couple of generations, has been well studied, and is understood relatively clearly – but the main effect has still been to compel the animal and human inhabitants to migrate southwards. In a post-Ice Age period a similar migration no doubt also happened – hence the spread of cattle herding into East and Southern Africa – though in these cases with the added complication in that central areas of the Sahara were left as islands within the accompanying desert. The inhabitants will therefore have become isolated in the still moist areas – at least that will have happened to many of the animals: it seems likely that the human inhabitants would decide to migrate, possibly taking some of the animals with them. One further development, therefore, is in the practice of herding, which was in effect a new version of nomadism. Whereas in India and Japan and Denmark the new warmer weather persuaded people to settle in one place for considerable periods of time, or to rotate through a few chosen sites where the hunting and gathering was rich and predictable, as on Oronsay, in the Sahara the principal means of survival for human beings was to gain control of a group of cattle, and then guide the animals to sources of food and water which the animals might never have found for themselves. No doubt there were favourite watering and grazing places which the nomads would visit periodically on a regular basis.

The earliest rock drawings imply the usual hunting life for humans. They drew the animals they hunted, just as their cousins in the European caves did several thousand years before, and just as the Indians, the Norwegians, and the Australians were doing at about the same time. But where the animals can only cope with the encroaching desert by migrating or dying, the people themselves could adapt. Those who stayed were, of course, much fewer in numbers, but they were able to change their lives so as to survive, and by selecting certain animals to bring under their control, they developed yet another successful response to the global warming of the end of the Ice Age. They were the ancestors of the modern desert dwellers and of the nomads of the desert edges.

The Saharans went through the same process of adaptation to the desert at the end of the 'Great Humid', as their Australian cousins had done much earlier. The two sets of people were faced with slightly different problems, but the essential difference was that in the Sahara there were 'islands' of highlands where water and food supplies for animals continued to exist when the surrounding lands became dry; the geographers' 'refuges'. It was within these 'islands' that domestication took place, because it was there that the animals could not escape, except to die. In Australia, these humid islands in the desert are less obvious, and the people and the animals could relatively easily get away to the coastlands, much of which were still humid. And, of course, in Australia there were not the biddable and domesticable animals available – no one could domesticate wallabies and kangaroos and the duck-billed platypuses. But the main difference was in the ability of Australians to continue their comfortable wandering lives, even in the great desert, whereas the Saharans were threatened with an early death if they did not do something drastic to enable them to find a way to survive.

Further Reading

For desertification consult Neil Roberts, *The Holocene, an Environmental History*, 2nd ed., Oxford 1998, particularly Chapter 4, and articles in *Tides of the Desert, Contributions to the Archaeology and Environmental History of Africa in Honour of Rudolf Kuper*, Koln 2002, by N. Petit-Marie, 'Large Interglacial Lakes in the Sahara-Arab Desert Belt', Andrew B. Smith, 'The Pastoral Landscape in Saharan Prehistory', and Angela B. Close, 'Sinai, Sahara, Sahel: the Introduction of Domestic Caprines in Africa'.

For Australia see the books by Josephine Flood noted in Chapter 1: also articles in *Antiquity* 'Transitions' supplement by Paul S.C. Tacon and Sally Brockwell on Arnhem Land, by Peter Veth on North-west Australia, and by J.M. Beaton on the coastal areas. Stanley Breedon and Belinda Wright, *Kakadu, Looking after the Country, the Gagudju Way*, is a well-illustrated account of that country

For the archaeology and rock art of the Sahara see Fabrizio Mori, *The Great Civilisations of the Ancient Sahara*, trans. B.D. Phillips, Rome 1998, and 'The Earliest Saharan Rock-engravings', *Antiquity* 48, 1978, 87–92; A. Muzzolini, 'Dating the Earliest Central Saharan Rock Art: Archaeological and Linguistic Data', in R. Friedman and B. Adams (eds), *The Followers of Horus, Studies Dedicated to Michael Allen Hoffman*, Oxford 1992, 147–154. Needless to say these two do not agree on dating or interpretation. The most convincing theory so far is that put forward by Fiona Marshall and Elizabeth Hildebrand, 'Cattle before Crops: the Beginnings of Food Production in Africa', *Journal of World Prehistory* 6(2), 2002, 99–143; for more detail on the crucial archaeology see Fred Wendorf and Romuald Schild, *Cattle Keepers of the Eastern Sahara, the Neolithic of Bir Kiseiba*, ed. A.E. Close, Delhi 1984.

Gallery VI

Rock Drawings from the Sahara

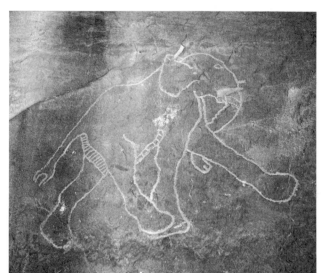

VI.1 A happy elephant. The art from the Sahara is in a variety of forms: drawings, carvings, paintings, but the essential thing is that they all tend to demonstrate that the Sahara was once populated by animals, which one would normally expect to find in a much wetter climate. Elephants (V.1) do appear fairly often, as do giraffes.

VI.2 A doodle of two giraffes.

VI.3 A giraffe hunt, one animal cut out and taken.

VI.4 Cattle, an early carving.

VI.5 A rock used frequently, showing a variety of artistic styles.

VI.6 Cattle, in a later, painted style.

VI.7 Men were not the only hunters; an ibex being pursued by dogs or perhaps wolves.

VI.8 People were a fairly frequent subject for the artists, depicted in a wide variety of actions.

Chapter 7

Agriculture

The Varieties of Response

T he effect on mankind of the increase in the global temperature turns out to be various and it certainly included the unexpected. In Europe the vicious Ice Age winter of the Last Glacial Maximum gave way – eventually – to a prolonged benign period of warm summers and cool winters, which permitted people in Denmark, for instance, and elsewhere to adapt themselves to a new life of near-sedentary foraging. The same effect was visible in Japan, though the sedentarism there was even more pronounced, and the foraging life there lasted much longer than in Europe. In India the people became equally fixed in place – or could be – but hunting and gathering remain the chosen lifestyle, perhaps because of the abundant game available in the newly moist subcontinent; foraging might be a reasonable description of Indian Mesolithic life in the Gangetic plain, where the inhabitants had such rich hunting possibilities that they appear to have been able to settle down in one place, but hunting the larger animals seems to have taken a larger part and to have lasted longer in time than elsewhere.

Desert life in Australia was perhaps ameliorated by a reduction in the overall size of the desert lands, but the flooding of the seaward plains crowded hunters into a smaller area, or forced some of them into adapting to life in a desert environment; the stress emerged as fighting and as a new reliance on foraging along the seashore – though hunting still remained the main source of food for most. In the Sahara the climate veered alarmingly between 'hyper-arid' and humid and back again several times over a lengthy period, so that the drying-out periods included the local domestication of food animals, and this was a development which helped sustain both the herders and the animals they herded.

In North America the melting of the ice allowed people living a hard Ice Age life in Alaska/Beringia to reach southwards into a more benign and productive area, an enormous land where hunting was good and the ice absent, while those they left behind adapted successfully to the cold climate which continued in the north – and all this from the small population in central Alaska in the last millennia of the Last Glacial Maximum; clearly an imaginative and resourceful group of people. In coastal areas elsewhere, such as South America, sedentary foragers concentrated on the fishing possibilities of the coast.

All these groups found that their lives were greatly improved by the adjustments they had to make because of the rising temperature. In some cases, the need to move repeatedly was reduced or even eliminated; in all cases the availability of food increased. Other groups preferred not to change if they could avoid it, and moved to new hunting grounds in the vicinity of the still cold regions – yet these were reactions in slightly different ways. Indeed, it was a built-in conservatism which is the most obvious characteristic of all those who responded in any way to the new climate. Those who moved did so to avoid having to change their lives: those who did not move changed their lives as little as possible. For the latter, in many cases they found that those changes they were compelled to make were actually beneficial, at least in the long term, though, of course, those changes then usually required still more changes until a new social and economic equilibrium was achieved.

The mammoth hunters who followed the mammoths when they went north did so in order to continue a life they clearly found productive and satisfying, and they and their descendants were then able to continue to enjoy it for several more millennia. Those who adopted a foraging life found that their food supply was satisfyingly various and abundant, and the exertions which were required of them were much less. The new pastoralists of the Sahara were able, once they had control of a domesticated herd of cattle, to organise their lives so as to ensure a constant availability of food, though they had to keep moving within a fairly restricted area to provide food and water and protection for the animals they herded. All of these groups were intent on ensuring their continued food supply, and all succeeding in doing so, but in different ways.

There was one further response made by hunters and gatherers to the climate crisis: the invention of farming. This phenomenon is the best-known response, for good reason, and it took place first of all in the Levant (or Near East, or Middle East) during the Ice Age, and so at the same time as the other groups mentioned in this book were puzzling out their own ways of staying alive as the great cold was ending. What is less well known is that this was both a development which took place several times in that region, but also in other parts of the world as well, all quite independently of what occurred in the Levant. All of these further cases occurred two millennia or more after the successful invention in the Levant, but they were all domesticating different crops, which usually required an individual and particular method of farming, and they were all at such a geographical distance from each other that in each case it was effectively an independent invention. It is just possible that it was the notion of farming which spread, and in some cases this is plausible, where later inventions looked for a local plant which could be domesticated and cultivated, where the original plant was unavailable – this certainly happened in at least two particular places much later than the basic invention in the Levant. But the geographical distance and time-distance between the basic domestications are such that it is best to assume independent invention. For the moment I shall concentrate on the original and first invention.

Farming in the Levant

Farming was no more likely or easy a development in food acquisition than any of the other more or less contemporary developments noted above – foraging, pastoralism, domestication of animals, and so on. What makes it different as an expedient which proved to be so unusually successful, of course, is the sheer productivity it revealed, which is one of those unexpected results, but was certainly not originally intended. Farming has conquered the world, and the competing developments – with the exception of pastoralism to some extent – have all given way before it.

The region in which farming was first used is that referred to on occasion as the 'Fertile Crescent', an archaeologist's term for a band of territory stretching from the mouths of the Tigris/Euphrates in Babylonia north-westwards along the valley of those rivers, west to the Mediterranean, and south to Palestine. The land involved in the development of farming

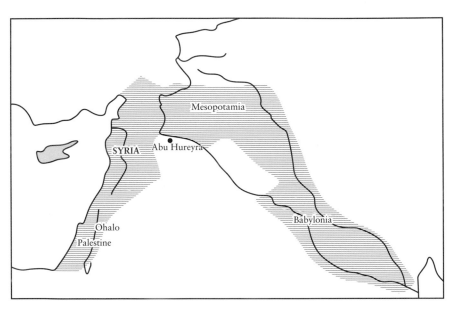

Fig. 53. THE FERTILE CRESCENT. The notion of a 'fertile crescent' of land in the Near East is an artefact of an archaeologist. In terms of early agriculture it is misleading since the early experiments and failures, and the eventual success, took place in Syria and Palestine, not in Babylonia. It was only when wheat and barley and rye and so on were successfully cultivated in the western part of this region that Babylonia was a possible agricultural area, where it became necessary to control and direct the waters to ensure growth.

was actually much smaller than this, and in particular it omitted the actual Tigris/Euphrates valley in Babylonia (Fig. 53), which, while later very productive, did not have the resources for its initial development. Agriculture only developed there after it had been invented in the north and west of the 'Crescent'. Babylonia was an area which was both wetter, in that the rivers flooded regularly, and drier, in that it was essentially a desert for much of the year. But it was possible to grow crops if water was brought to the dry land by irrigation. It was one of the crucial discoveries about the domesticated crops of the early farmers, that those crops could be persuaded to grow in regions outside the original wild crops' home.

The northern and western parts of the Fertile Crescent were rather moister than now in the period following the Ice Age, and rather drier earlier – the climatic changes which came with the end of the Ice Age were erratic and the quantity of moisture tended to fluctuate over fairly lengthy periods. The snow line on the mountains was 500 to 1,000 metres lower than now during the Last Glacial Maximum, but the desert areas of Arabia and Syria were probably just as extremely dry as now, so the 'Fertile Crescent' was a good deal narrower than the usual definition, cooler and dryer, and with the deserts encroaching into the modern steppeland areas it could not be considered particularly fertile. During the Ice Age the general vegetation of the moist area – Syria/Palestine and Southeast Anatolia – was open woodland shading to desert by way of steppe. This last area was a home to animals, especially caprines (goats/ sheep) and gazelles, but the woodland thickened up and the steppeland drove back the desert edge during the moister post-glacial times.

The area under consideration here is less than the full Crescent. It consists of the modern countries of Syria, Lebanon, Israel, Palestine, and Jordan, the lands facing the Mediterranean, together with an extension east to the upper Tigris area, the original Mesopotamia ('land between the rivers'). The most convenient term for the whole is the Levant. It is a very varied land, including the high mountains of the Lebanon and Anti-Lebanon, bounded on the north by the Taurus Mountains; down through Syria and Lebanon and into Palestine/Jordan runs the deep trench of the Rift Valley and the Dead Sea depression. The land has, or had, perennial rivers and swamps, and, of course, on the west it has access to the sea. It receives its moisture from the west, from the Mediterranean, but the area which is moistened by rain is narrow, for the mountains intercept much of the rain, leaving a rain shadow to the east, so that the land shades quickly to desert behind the mountains – with the exception of 'Mesopotamia', where the rains penetrate through a lower mountain barrier in north Syria to reach well to the east as far as the Zagros Mountains of Iran.

This land had been inhabited all through the Ice Age by hunters and gatherers. Many of the larger animals known elsewhere seem to have been absent, but excavations in caves on Mount Carmel in Palestine, Ksar Atil to the north in Lebanon, and inland in Palestine at Umm Qatafa all indicate that the normal fauna of the time included boars, red and roe and fallow deer, cattle, goats, horses, bears, gazelles, foxes and several smaller animals. (The country does seem to have had elephants later, and the usual predators – besides man – including lions.) This was the suite of animals whose descendants moved north into Europe with the retreat of the ice. All of them were suitable prey for hunters, and many of them remained in the Levant. The climate was suitable for supporting a considerable population of these animals, and their presence implies the existence of open woodland where there was some moisture, at least close to the coast and the mountains, but there was rather greater area of desert than at present, and there was a wide area of steppeland suitable for grazing between these two regions.

The climatic changes which ended the Ice Age were as erratic here as in, for example, the Sahara. The cold and dry climate of the Last Glacial Maximum gave way to the warmer phase of the Allerod interstadial, beginning about 15,000 years ago, in which the general temperature rose by about 4°C, a time which in the Levant was also wetter than before. This was reversed about 13,000 years ago, with the climate reverting to the cold and dry 'Younger Dryas' phase; then about 11,500 years ago the warming resumed, and this time the general temperature gradually increased by about 7°C, and finally stabilised into the present miscalled 'optimum' after a couple of millennia.

The Grasses

In this region there were, surviving in the cold periods in ecological 'refuges', primitive wild versions of wheat, barley, and rye grasses. As the climate warmed the rainfall increased and these plants were able to expand their range throughout the Fertile Crescent and beyond. They inhabited the open steppe which lay between the coastal woodland and the desert along a long stretch of country from southern Palestine to western Iran, and also in a similarly long narrow part of south central Anatolia, north of the Taurus Mountains. These grasses – emmer and einkorn and barley – are the ancestors of today's food plants. Emmer grew in the Levant, and einkorn probably in southern Anatolia, both at first in fairly restricted areas; barley, a hardier plant, grew more widely in these areas and spread further afield in western Anatolia, North Africa, and eastwards as far as Afghanistan. In addition, these same areas, notably north Syria and southern Anatolia, were the homes of the wild original versions of peas, lentils, and chickpeas, which are also plants which were to be domesticated early.

It was the grasses, emmer, einkorn and barley, oats and rye, which became the most important in the evolution or invention of agriculture. These plants all have certain characteristics which were important for their domestication and development. They were self-seeding, which means that once they were domesticated they did not become constantly crossbred with their less developed and still wild cousins. They were, of course, adapted to the Mediterranean climate of the Levant in that they germinated and grew quickly during the wet winter season. Their seeds were scattered at ripening, and being protected by a tough outer casing, those seeds then lay dormant through the hot dry summers until the rains returned, when the germination began once more. When humans began to harvest such crops they learned to do so just before the 'shattering', as this method of seed dispersal is called. They had to dry the seeds artificially, either by laying them out in the sun, or by roasting them, and then they had to grind the seeds to extract the food, which was a very small part of the whole. In doing all this, of course, they learned a good deal about the plants in practical terms, knowledge which could be used in domesticating both these and other plants. But one of the more difficult questions to answer is to wonder how humans learnt to do all these processes; they are surely not obvious. Was it accident, or deliberate experimentation?

In essence, the gathering of the harvest of the wild plants was originally a type of foraging, and so it fitted in well with the original life of hunting and gathering, though their nomadic life had to adapt somewhat to cope with the behaviour of the plants. The harvesters had to be present where the plants grew when they were almost ready to disperse their seeds, and for a time during and after that time in order to collect and process the harvest. Like the foragers in Japan about the same time, they were therefore pinned down in one place for a considerable time, harvesting, drying, grinding, preparing the food. This is not necessarily very much of a behaviour change for the people, but in order to secure a future harvest at that place, some seeds had to be left to germinate during the next winter, and some would need to be deliberately planted. This was a decision obviously difficult for hunters and gatherers, whose instincts would be to consume any food as soon as possible. Storage of food had undoubtedly taken place during the Ice Age in winter in frozen areas, but the heat of the Levant summer was not the place to attempt to preserve any fresh food. Having ensured that the seeds for the next year were planted, it will then have been necessary to guard the crop, both from other human foragers and from grazing animals. Once the decision had been made to rely on the grasses, at least some of the people had to adopt a largely sedentary life.

Ohalo (Fig. 54)

This change of life appears to have taken place more than once in the Levantine homeland of the plants involved, but on at least two occasions subsequent changes in the climate rendered the change null. (There may have been other attempts.) The earliest example so far located of humans relying to some extent on these grasses is at a site called Ohalo, in the Jordan Valley. The people of this settlement – it was a small village, in effect, with huts and hearths – lived beside an enlarged Sea of Galilee about 19,000 years ago – before the Last Glacial Maximum got under way seriously, therefore. This early reliance on food from grasses apparently ended with the intensification of the cold and dry climate which set in with the Last Glacial Maximum.

It is worth noting that, beyond planting and reaping, a series of other processes were developed to render this food palatable. The product can, of course, be consumed directly out of the seed, where each seed contains a small edible quantity which is rather chewy, but not very pleasant to eat. Collecting it all together, drying it, then grinding it produced flour, which could then be baked into bread, of a sort; these are all processes which had to be invented by the farmers, or perhaps by their wives. It is not clear if all this was known to the people of Ohalo, but it certainly was to later incipient farmers, and a considerable period of experimentation probably lay behind the time between Ohalo and these later sites.

In the same area as Ohalo, but perhaps from four millennia later, a number of sites, caves and camp sites in the southern Levant have produced other signs of a reliance on these grasses. The Ohalo people had used sickles made of microliths set in wooden hafts to reap their crop, which implies at least harvesting on a regular basis and the invention of a specialised tool for the task – and the processes of preserving and

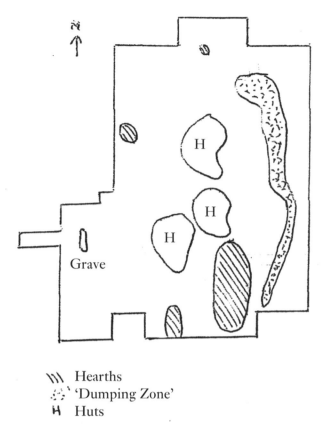

》\\ Hearths
∴ 'Dumping Zone'
Ӊ Huts

Fig. 54. OHALO, THE ORIGIN OF AGRICULTURE. The plan of the excavated area of Ohalo village. The layout indicates care: the huts are grouped together, rubbish is to be thrown away to the east, hearths are kept away from the huts (which were made of brushwood). The grave is close to the village so that the dead remain part of the community.

then deliberately planting the seed were no doubt also undertaken. The second group of sites, in a culture called the 'Geometric Kebaran' by archaeologists, did not used sickles, it seems, but they did have mortars and pestles in which to grind the corn, and again this implies that regular harvesting and the grinding of the crop had taken place, again with a specialised tool invented for the job. They were thus producing flour, and probably baking it as well.

Both of these groups of people clearly foraged for their crops and used them for food, but it is difficult to see any evidence of actual farming at any of these sites other than the fact that they cut the grasses at harvest time (which means that they were conducting the harvest before the shattering took place, in order to make sure they collected the seeds, which would also then require threshing and drying. The evidence of such work would include storage facilities, digging tools, perhaps hoes, and of course some means of baking. That is, in the sense of deliberately planting the seeds, and the cultivation of the plants, these people were not yet farmers. The final clincher is that the plant remains found at the sites are always of the wild varieties.

Nevertheless, the harvesting of these foods was a successful strategy for human survival. The Geometric Kebaran sites are relatively numerous, and are spread through the Levant from Palestine to north Syria. The culture developed later into the Natufian, a culture which spread still more widely throughout the Levant, with a denser pattern of human occupation than ever, either there or elsewhere. Reliance on farmed food had therefore developed, and the settlements are sufficiently numerous and concentrated to suggest that hunting was now no more than a residual food resource.

These Natufian settlements, like Ohalo, were essentially small villages consisting of a few houses, but probably twice the size or more of any preceding human settlement. Thus even the most primitive agriculture compelled the invention of a series of new tools and processes and skills, and brought a major change in human behaviour. There are also other similar sites which were not technically or archaeologically Natufian in culture, spreading into southern Anatolia, and which operated the same agricultural economy. It would seem therefore that the Natufian practice of grass foraging was soon copied by other groups who kept their own culture – unless they developed the practice for themselves. It seems more probable that the idea and the practices spread through example and imitation.

By about 13,000 years ago such village sites were spread throughout the Levant and into southern Anatolia and as far as western Iran – which is just the area where the wild cereals and legumes themselves had spread from their 'refuges' during the warmer climate after the rise in temperature. Of course, this life may have been adopted because of a shortage of other foods – farming was, after all, a laborious business.

These first agricultural experiments, if that is what they were, were essentially foraging, in much the same way as the proto-Danes regularly harvested their shellfish from the same areas, or the Oronsay foragers moved about their island with the seasons. The Geometric Kebaran culture has been interpreted as that of people who migrated into lowland areas in the winter, presumably to protect and supervise the growing crops and then harvest them, and then moved back into the hills in the heat of summer. The gathering of the grass crop was only a part of their food resources, for hunting was also a large part of it, as was the gathering of other foods, such as nuts and fruits, especially in the periods before the harvest of grasses was ready. It was a life which was clearly fairly successful, judging by its widespread adoption, and reasonably well adapted to the fairly uncertain climate of the time in the region.

The rise in temperature at the end of the Last Glacial Maximum during the Allerod interstadial was what had enabled the plants which the Natufians and the Geometric Kebaran people harvested to spread out of their 'refuges' throughout the Levant, but it was also a time of considerable climatic fluctuation, in part the result of such events as the widespread flooding of coastal lowlands (as in Sahul, the North Sea, the Gulf, and parts of the Mediterranean) and of the meltwater pulses of the great North American ice-trapped lakes, a time of climatic turbulence which lasted to the end of the Younger Dryas. All this fluctuation must have had its effect, either to encourage or retard the invention of farming's processes, but once begun it was a technique of food production which was evidently worth pursuing.

So far, this was only a partial response to the global warming, since the erratic changes in the climate meant that this incipient agriculture was liable, as in the earliest example of Ohalo, to fade away in a period of changing climate. It can be seen in fact that the Ohalo settlement was primitive in a sense that the plants harvested there were probably an isolated refuge group, and so it was not possible at the time to extend that farming lifestyle outwards beyond the small area around the site (though the dating of the Ohalo settlement puts it in a warmer period, before the greater cold of the Last Glacial Maximum.) It was, in that sense, a false start, and the renewed cold killed it, though there may have been other such 'false starts', and the idea of gathering grass seeds for food – not the most obvious procedure – was clearly in the air in the Levant from at least the time of Ohalo onwards. The second attempt – the second we know about – was the Geometric Kebaran leading into the Natufian, but this time the process was not seriously interrupted by a drastic climate change, though it was a precarious development nevertheless, thanks to the cold period of the Younger Dryas, which lasted well over a thousand years. This is emphasised by the fate of the first place where the next step was taken.

Abu Hureyra

Abu Hureyra, in the valley of the Syrian Euphrates, was a largish village which developed about 13,000 years ago (during the Allerod interstadial warm period). The livelihood of the people was based on the cultivation of rye. That is, the people were practising true farming. They were deliberately planting part of their harvested seeds instead of merely leaving some in on the ground as the other Natufian proto-farmers had done. However, as at Ohalo, this was still a fairly primitive practice (Fig. 55).

The Hureyrans, like the Natufians, and like those at Ohalo, were harvesting wild plants. To take the next step it was necessary to select the seeds to be replanted, concentrating on those plants where the seeds did not 'shatter' at dispersal, for by harvesting all the seeds just before that event no selection was taking place.

The proportion of non-shattering seeds increased by the very harvesting method used, since the non-shattering seeds were more easily collected, and the shattering variety more easily lost. This clearly took a long time to have an effect on the quality of the crop, but a long period of familiarity with the needs of the plants led to the selection of non-shattering seeds becoming automatic. None of these early farming attempts succeeded in deliberately selecting the non-shattering seeds, which would be extremely laborious and time-consuming, but they surely appreciated the change which was gradually coming about in their harvest.

The settlement at Abu Hureyra is usually described as a village, or a large village, but it contained a larger population than any other human

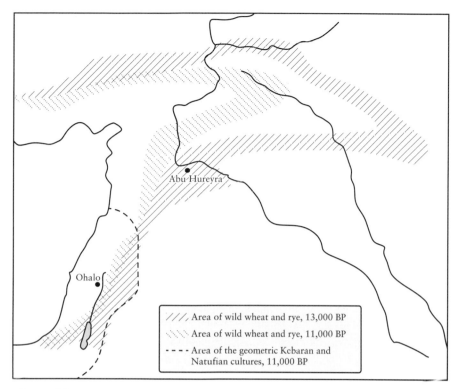

Fig. 55. THE FIRST ARGRICULTURALISTS. The area in which wild cereals were native shifted with the climate changes. The shift was clearly a disaster for Abu Hureyra. To the south the Geometric Kebaran and its successor the Natufian cultures covered a much larger territory, and so overlapped both the earlier and the later ranges of the wild crops. This is one of the reasons for their success where Abu Hureyra failed.

settlement anywhere until that time. The population has been calculated at several hundred, perhaps 400 or 500. This ranks it almost as a town – some mediaeval European 'cities' were smaller than this – and it is sufficiently large that we must imagine the invention of some sort of government to keep order, and to regulate and organise defence, and perhaps to organise planting and harvesting. This was a rich place, when compared with any other nearby place or group, and will have excited envy.

What seems to have pushed the people of the village at Abu Hureyra into the direct cultivation of grass crops, as opposed to the harvesting of wild crops, was the beginning of the renewed cold of the Younger Dryas, which began about 13,000 years ago and gradually intensified, reducing rainfall and the moisture in the air. Abu Hureyra is in territory which had become partly wooded during the preceding preliminary warm Allerod period, but as the cold grew, the woodland died back. Abu Hureyra was therefore left in a dry area of steppeland. No doubt the disappearance of the woodland reduced their opportunities for hunting for meat as well as for gathering fruits and nuts and other forest products. They attempted to compensate for this reduction in their accessible resources by increasing their reliance on the grass crop. The experiment was only briefly successful. The evidence is that the rye they cultivated was a variety bred specifically for cultivation, and the weeds which infested the fields were those which always turn up in cultivated ground – a hoe was another necessary invention. But rye is hardly a grass crop with much in nutritional value, and its use implies that the land had become considerably too cold for wheat or barley, the two most productive of the grasses, to flourish. In the end the people eventually abandoned their village in the face of the difficulties; there is considerable evidence of hardship and malnutrition in the skeletons which were left in the village cemetery. Clearly they held on as long as possible before fleeing to a more productive area, presumably to the north, or perhaps towards the coast. The later inhabitants of the region herded sheep and goats, but the human population was clearly much fewer.

Natufians

The relaxation of the cold at the end of the Younger Dryas period, from about 11,500 years ago, encouraged the growth and cultivation of the harvesting grasses once more. But in order to reach the stage of cultivating domesticated grasses it was necessary for the farmers to learn to select the seeds to be replanted – that is, a communal decision had to be made to move from the harvesting and consumption of wild seeds to the deliberate planting and harvesting of fields of crops, and then storing the seed corn for the next season. The planted fields had then to be protected from wild grazing animals, and from the farmers' own animals where they had them, and when the harvest came the whole community had to turn out and gather the crop as rapidly as possible. By about 10,500 years ago it seems that the deliberate selection of non-shattering seeds, and the cultivation of domesticated plants definitively replaced the casual

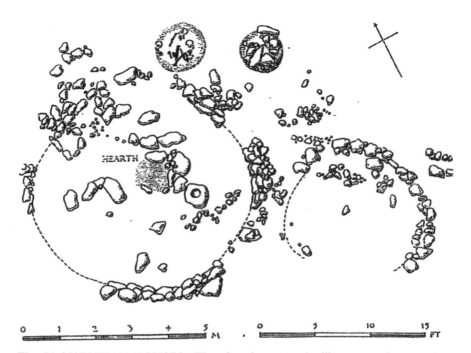

Fig. 56. NATUFIAN HOUSES. The plan shows two dwellings – or a house and an annexed hut – of the Natufian period at Eynan in Palestine (compare with Ohalo, Fig. 54). The houses are circular, with a ring of stones forming the foundation of a superstructure of wood and branches. A hearth is central in the main hut, and next to it is a mortar for grinding grain. Two graves are close by, so that the community, past and present (and future, in the children) remained together. This is intended as a permanent settlement.

harvesting of wild plants as the main source of food for the communities (Fig. 56).

This is what makes the Natufians distinctive. It will have been gradually realised that the non-shattering plants could be harvested separately, since the seeds did not disperse at once, and so the proportion of them in the seed collected and preserved for the next planting would be gradually enlarged. No doubt the extra task was not a popular one, at least until the results became clear.

It will be seen that the conversion of hunters and gatherers into farmers was neither a quick nor an easy process. Moving from the earliest known case of harvesting to the fully domesticated farming of grass crops – from Ohalo to Natuf – took up to ten millennia. During that time there were at least three false starts (Ohalo, the early Natufians, Abu Hureyra), and only after several millennia of foragers reaping the wild grains did farming – planting, caring, weeding, harvesting – emerge as a full-time human activity; however, by about 10,000 years ago, this had happened.

Gobekli

About fifty kilometres north of Abu Hureyra is the site of the curious structures of Gobekli Tepe, twenty of them; there are also half a dozen other places which contain similar structures in the same, fairly restricted, region. The distinctive buildings are circular walled areas, sunk into the ground, each of which contains a number of T-shaped pillars, some with low relief carvings of animals or birds or insects on them. The date of the origin of the structures, according to radiocarbon dating, is about 11,000 years ago, in a period called by archaeologists the Pre-Pottery Neolithic. This is approximately the same date as when the Natufian farmers of Palestine were selecting non-shattering seeds in their fields.

The Gobekli structures are enigmatic, and have therefore been labelled as a 'temple', though there is little or nothing to suggest that they were religious in purpose. This remains a possibility, but other interpretations should not be ruled out. Later in Syria, for example, there are similar slightly sunken circular buildings which are clearly secular meeting places. They have benches around the inside walls, except at the doors; the buildings were roofed. At the Gobekli site, four of the circular structures have been excavated, and have proved to include stone benches. The whole may have been roofed; two of the pillars are taller than the others, which could be the roof supports, the roofs probably consisting of wooden beams and thatch. It looks to me more likely to be a meeting place than a place of worship (though it could be both, of course); it is all too easy to fall back on a religious classification when an interpretation of a building is not obvious, and it is all too therefore to be wrong. The animals and so on which are carved on many of the pillars may be clan symbols, or they may simply be pictures (Gallery VII).

Little evidence of the economy has been found, but it is assumed that the builders inhabited villages nearby (these have not been located either), at least for part of the year. The date of the origin of the structures is exactly the time when agriculture was beginning in the region, after Abu Hureyra's failure, and is roughly contemporary with the Natufians – that is, the development of agriculture was a local activity in all the lands from Palestine to Anatolia. Given the effort and the time involved in excavating, carving, and building the structures it is almost compulsory to accept that the people had an agricultural economy, in which time would be available for the work of building, and that they were therefore essentially sedentary. It is also, given the other sites with similar structures in the region, a society which had spread over a considerable area. It has been generally assumed that this society was one of hunters and gatherers, but this looks most unlikely given the requirements of the production of food to provide the energy for the work of building. The dating puts the origins of the society's efforts into the Younger Dryas period, when there was pressure on resources for all the incipient agriculturalists in the region. It is significant that the whole system of walls and pillars were deliberately closed down and the structures filled in with refuse about 10,000 years ago. No doubt this was done because the structures, whatever they were for, were no longer needed, but possibly because the warmer, moister climate enabled the population to spread out over a larger area – though a religious explanation has been adduced for this as well.

Animals

Plant domestication is only a part of the process of the invention of farming. The domestication of farm animals is the other part. This is something which took just as long to accomplish. The Natufians and the Geometric Kebaran peoples had relied as much on hunting for their food as on harvesting the grasses, and indeed quite likely more so, but they had also to protect their plantings and the planted areas, from their animals, and from the wild herbivores. Herding of the more biddable animals is hinted at in the near desert areas, such as Sinai and east of the Jordan, well back in the Last Glacial Maximum. The animals involved were goats and gazelles, and the evidence is largely that at some sites such as Beidha, not far from Petra in modern Jordan, the proportion of these animals' bones is very high. But this is little more than a version, once again, of foraging, just as the early harvesting of grasses was – the two were clearly contemporary. It was alternatively perhaps a matter of driving the animals, rather than herding them.

The goats and gazelles were the first animals to be herded, or driven, no doubt, because they were animals of the dry lands, and were therefore all the more valuable to the hunters. Within the moister lands the animals which proved to be domesticable were sheep and cattle and pigs. Sheep may well be a domesticated variety of a type of goat, having been bred into submission and timidity by selective breeding, but neither cattle nor pigs were at all biddable in the wild state – who would tackle a bull or a boar with the intention of taming it? But these animals did feed on grasses, and their young could be collected when the adults were driven away from the fields.

These animals seem to have been fairly scarce in the southern Levant, but more common in the north, in modern Syria, northern Iraq, and southern Anatolia. (They may have been part of the resources exploited by the people at Gobekli.) At least it is in this northern area that the earliest evidence for domesticated animals is found, notably at Cayonu in south-eastern Anatolia. There the remains of domesticated goats, sheep and pigs have been found in a context dating to about 10,500 years ago – cattle evidently took a good deal longer to bring under control (though that was happening at the same time in the Sahara). This is just about the end of the Younger Dryas cold period, and it seems reasonable to assume that the drier, more difficult climate of the period had, as in the Sahara, brought men to see these animals less as wild prey and more as a resource to be conserved.

It will not have taken the new farmers long to discover the value of these animals for manuring their fields. The practice of combining animal and plant farming spread fairly rapidly throughout the Levant and into Anatolia and eastwards into Iran. The Iranian version emphasised cattle, whereas the Levantine version emphasised wheat and barley. Nevertheless, the whole set of plants and animals comprised the farmers' kitbag, and it was this set of items which was spread. To develop the whole suite of plants and animals and skills had taken well over ten millennia, with several interruptions on the way, and only in the last millennium or so at the end of the Younger Dryas had it all come together as a balanced system of economy.

All this had taken place before the Levantine peoples began to use pottery – the period archaeologists call the Pre-Pottery Neolithic ('Neolithic' is the term now signifying the time of agriculture). This was long after pots had become a familiar product in eastern Asia. The several millennia separating its invention in the Far East from its adoption in the Levant might suggest that its use spread outwards from a single centre of invention. There is, however, no signs of its transmission across Asia – no case of pottery as early as this is known in Central Asia or India, for example – and we may suppose, as a preliminary theory, that it was invented anew in the Levant. As in eastern Asia, the use of pots was especially helpful to those whose food was derived from farming (and foraging), since it allowed the preparation of such foods as porridges and other semi-liquid preparations, as alternatives to the tougher foods they had to eat before; it was also a useful tool for storing food, away from insects and mice.

Spread of Agriculture

The Levant is reckoned to be the area where the earliest domestications took place because of a detailed and convincing series of archaeological investigations in the area. But it may be that there were other regions where similar domestications by other peoples took place. After all, it happened several times in the Levant; it could surely have happened once or more elsewhere. It is likely, for instance, that the domestication of animals had happened in the Sahara before the Levant (Chapter 6). Perhaps other domestications had happened at the same time, or possibly even earlier. These other regions have not been so well investigated as the Levant so far, but it is clear that independent domestication did take place in other regions, using different plants and animals. (In a way the discovery of plant domestication and the invention of farming in the Levant was one of those events which came about as a result of looking for something else – the early excavators were searching for evidence of biblical events and history, not general history; in the same way the miasma theory was aimed at the elimination of malaria: that is, research is worth pursuing for its own sake, and its revelations are not always what was looked for.)

Once the techniques of farming had been developed, they were easily transferable from one society to another, or one region to another, as with the adoption of foraging for grasses transferring from the Natufians of Palestine to people of a different cultural tradition in Anatolia, though the crops were the same and the types of country they all lived in were very similar. So two methods of transfer are possible: one, the direct transfer of the techniques of farming the same crops and animals, and, two, the transfer of the *idea* of farming, which would not necessarily involve the cultivation of the same crop or crops, but would involve the domestication of other plants, perhaps crops which were more congenial to a different climate. However, when we can see the domestication of different plants taking place at a great geographical distance from the Levant and its emmer and einkorn, wheat and barley and rye, sheep and goats, it is very likely to be a completely separate and independent development, a similar reaction to a similar problem.

Rice Farming

One significant region where such a separate domestication took place was China, where the plant involved was rice, whose domestication, it seems clear, took place independently of that of wheat and barley. This was thus the place where it can be argued that it was an example of the second sort of domestication, where the idea was adopted, and a different crop domesticated, though as steadily earlier evidence of rice domestication emerges, independent invention seems increasingly probable. It would be best, therefore, to see the development of rice farming as an independent work.

The region where the early development of rice cultivation occurred is in the centre of the country, in the middle valley of the Yangzi River and in the nearby valley of the Huai River which joins the Yangzi from the north (Fig. 57). The details are not yet altogether clear, and the process remains largely a matter of guesswork and assumptions, but one site in particular illustrates the stages of the process by which rice domestication took place. This is a cave at Diaotonghuan in Jianxi province, some way to the south of the middle course of the Yangzi, where excavations have discovered a sequence of stratified occupations containing a whole series of remains of rice. The earliest record begins about 13,000 years ago (the beginning of the Younger Dryas cold period). This was with wild rice, which at that time formed over 90% of the finds of rice; less than 10%, therefore, was domestic rice, which could in fact be included even in a wild grown crop. This is, of course, the non-shattering type, which is a natural mutation of the plant, as it is with wheat and other grasses. The earliest level in the Diaotonghuan series is therefore not evidence for the deliberate cultivation of the crop, but only of foraging for the wild variety, which naturally included some of the non-shattering types – the equivalent of Ohalo foraging in the Levant (Fig. 58).

That first period of rice consumption by the people of the cave was followed by a gap in the cave's use for about three millennia or so. When people again came to inhabit it, about 10,000 years ago, they were able to harvest rice, and the domestic non-shattering, type formed a third of the total. This increase could have been achieved simply by the selection of the seeds by casual harvesting, and not necessarily by deliberate planting, for the mere practice of harvesting would increase the proportion, since, as with wheat and barley, this domestic variety was easier to harvest, and less was lost. But to promote such selection, even if accidental, does imply the deliberate planning of a seed crop saved from the previous year's harvest. The intervening three millennia, when there were no rice finds from the cave, coincided with the return of the cold climate of the Younger Dryas, the event which so disrupted agricultural development in the Levant, notably at Abu Hureyra. In China, however, it does not seem that the use of grass crops had progressed so far as in the Levant. The earliest level at the Diaotonghuan cave was contemporary with Natuf, where the deliberate use of non-shattering wheat and barley was already clear.

The subsequent gap in the Chinese record suggests there was a long delay in further progress, at least in this area of China, consequent on the new cold period. At the same time, the increase in the proportion of non-shattering rice over that period of three millennia does show that rice harvesting continued elsewhere, presumably somewhere to the south of the Yangzi where the climate permitted wild rice to grow and to be harvested. Repeatedly harvesting the same fields will steadily increase the proportion of non-shattering seeds, since the normal wild type will

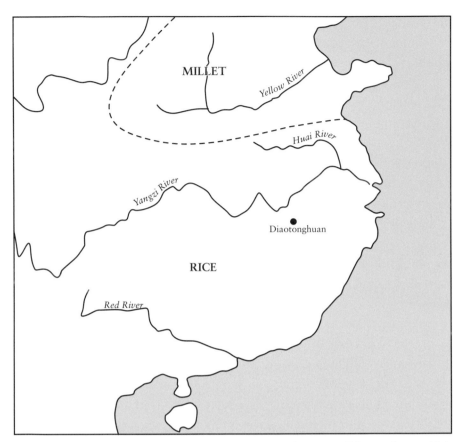

Fig. 57. EARLY AGRICLTURE IN EAST ASIA. The cave at Diaotonghuan which has produced good archaeological evidence for rice domestication is on the very northern edge of the area where wild rice grew. To the north, beyond the hills and the Huai River valley, is the millet region. As the climate warmed after the Younger Dryas cold spell, rice farming became practical in the Yangzi and Huai valleys, and the idea of farming was soon taken up further north, where millet was the crop.

more easily fail to be harvested while non-shattering plants would always be collected.

The area where the Diaotonghuan cave is situated is at the northern edge of the range where wild rice grows as it exists today, and as it probably also was in the pre-Younger Dryas period. The resumed cold will have driven the boundary of the plant's range southwards, away from the cave, whose inhabitants were therefore compelled to change their diet, or at least to drop rice from their foraging activities, if they stayed in the area, though it seems they also moved away. (This was, of course, the same set of choices faced by other communities in a time of climate change.) The renewed warmth at the end of the Younger Dryas will have brought the wild rice back, after a time. The increase in the proportion of the harvest which is of the non-shattering 'domestic' type suggests that the seeds for the crop might have been acquired from another farming group, or the new inhabitants brought the seed stock with them. This is, however, by no means certain, but it clearly implies a community which was cultivating rice, and was spreading northwards to colonise new lands; this is an area of research in which new discoveries may be expected.

This was a region which already used pottery vessels (as noted earlier in Chapter 5). The development of the preparation of rice for eating requires the use of such vessels, even more than is the case with wheat. The wild rice was no doubt prepared by grinding, but preparation by boiling was soon adopted, as being easier and providing a better food. It may be that the domestication process was as dependent on the availability of pots as of the rice itself. Similarly, the most productive method of growing rice turned out to be in flooded fields, and this would have taken time to discover and to organise.

The Younger Dryas cold period did not, of course, involve the loss of the technique of pottery making as it did the loss of the rice crop, and wild rice was still growing in lands to the south of the Yangzi valley; evidence for its cultivation and harvesting there is not yet available, though it may perhaps be presumed. Over that cold period the foragers gradually increased the percentage of non-shattering rice, possibly deliberately, so that when rice collection returned to Diaotonghuan sometime over 10,000 years ago, it cannot have taken long for it to be realised that the non-shattering variety could be selected and stored, perhaps first as a reserve food supply, and then separated out during the harvesting for replanting in the next spring. By the time another millennium had elapsed (so about 9,000 years ago) the percentage of 'domestic' rice in the Diaotonghuan cave was half the crop and perhaps more, and so by then one could say that deliberate rice farming had clearly begun: the wild rice was still being foraged as a supplement, simply because wild rice seeds were included in the previous year's harvest, but the community now

RICE AT DIAOTONGHUAN CAVE

A. Cave stratigraphy (diagrammatic)

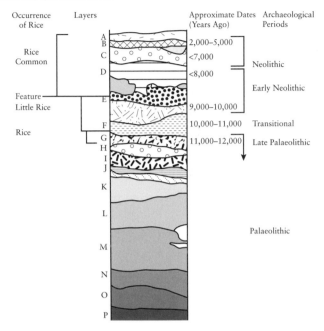

B. Presence of Rice (%)

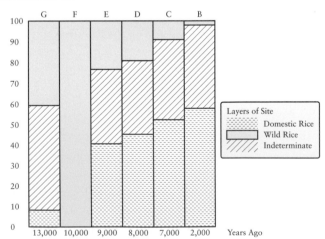

Fig. 58 RICE AT DIAOTONGHUAN CAVE. The cave had deep and continuous stratigraphy, the changes dated by the artefacts. The layers in which rice was common contained pottery of increasing technical accomplishment; occasional traces of rice were found in most strata; only from 12,000 years ago onwards were the remains substantial. The rice phytoliths could be classified as wild (this included all earlier than 12,000 years ago), or domestic (i.e., cultivated) and an indeterminate type.

relied mainly on the cultivation of the domestic non-shattering rice for its sustenance.

This process cannot yet be precisely detailed from the Chinese evidence, but what can be understood as having happened is sufficiently similar to what occurred in the Levant to allow the better understood process of domestication of other plants in the Levant area to stand in for the missing elements of the rice story. The basic conundrum lies in the issue of how deliberate was the selection of the non-shattering wheat and barley and rye and rice. Some non-shattering plants naturally existed in each crop, and no doubt the foragers would have come to recognise the type, for they were surely intimately familiar with their plants. But the continued harvesting of wild rice for a good millennium after the non-shattering variety became fairly common in the diets of the foragers, both in the Levant and in China, even as the percentage of wild as against non-shattering seed declined, implies that it was only over a very long period that the non-shattering type became recognised as the one it was best to plant. It was this type which was easiest to harvest, for it can be left to be cut when ripe, whereas the shattering type had to be collected while still unripe or it would be lost, and then it must be dried out artificially; this last technique was not always necessary with the domesticated plants, since it is possible to wait to harvest them when they were fully ripe. What can be said is that in both the Levant and China the recognition was ultimately made, and it was at that time that real farming, in the sense of people deliberately planting, cultivating, and harvesting controlled domestic crops, began.

Millet

In the Levant there is evidence that the techniques of harvesting and farming barley and rye could be passed on to communities of different traditions, from Syria to Anatolia and Iran; in China there is similar evidence, but this time it is clearly a transfer of the idea of farming and domestication which is happening, not the techniques of rice farming; the transfer was to people who then domesticated a different plant. To the north of the rice area, in the Huanghe (Yellow River) Valley, wild rice was only an occasional plant. Instead the locally foraged plant was millet.

There were two wild versions, foxtail millet and broomcorn millet, which were widespread throughout the cool steppelands from northern China as far as Eastern Europe, and no doubt it was normally consumed in all that area. But it was in China that the plant became domesticated and cultivated. The earliest date for its cultivation which is so far known, at 8500 years ago, is rather later than the revival of rice harvesting in the region of the Diaotonghuan Cave – that is, rice farming had reached as far north as the edge at least of the millet area; the two plants overlap in their wild ranges. Again the process of its domestication is invisible to us, at least in the state of our present knowledge, but it seems very likely that the practices of rice domestication and cultivation became known in the millet area, and it then becomes very likely that the cultivation of millet was deliberately adopted by people who were familiar with rice farming but were unable to grow rice in the north because the climate was not suitable; millet was therefore turned to as a suitable local alternative. That is, it was probably a deliberate decision to take up the cultivation of that local alternative to rice, which in turn would have required a good understanding of what was required in order to domesticate a crop.

(It is worth noting here that this adoption process is one which is quite normal in economic affairs. Economic history is littered with parallel examples by which a technique is pioneered in one region, where the inventor makes repeated errors and mistakes and gallops off down unproductive side-tracks in the development until after a long struggle the finished and efficient method of production is perfected; it is then copied in other areas without the need for them to go through the painful experimentation and developmental period. The spread of mobile phones into new areas, into some of which the earlier period of telephone lines and exchanges and hand-held sets had not yet reached, is only the most recent example of this, but it is a process with a long history; the pioneering of industrial techniques in Britain in the eighteenth and nineteenth centuries and their subsequent adoption by competitors is a rather more substantial case, as is present problems of the pharmaceutical companies with the expensive development of new drugs followed by non-licensed production elsewhere. In China and the Levant the extension of agriculture to new areas, new crops, and different peoples, shows the same process at work at the end of the Ice Age.)

Despite the probability of the transfer of the idea of the cultivation of domesticated plants from the rice areas to the millet areas in China, and the techniques being transferred from the Levant to Anatolia, it seems highly unlikely that either the idea of domestication or the techniques of farming travelled from the Levant to China. The process of the domestication of wheat and barley and rye in the Levant was a long drawn out, hit or miss process, one which was always subject to interruption by the vagaries of the post-Ice Age climate, and was not successful until sometime after 11,000 years ago, if one assumes that the Ohalo site is just about the earliest case of successful concentrated foraging for these crops and the Abu Hureyra site was another – but both of these failed and died off in the renewed cold; the post-Younger Dryas spread of the technique was its eventual success, but only after several such failures. By that time, as the Diaotonghuan Cave exploration shows, China was already in the process of domesticating rice with some success, but was going through a very similar hit-and-miss experience in doing so. When that process began, success in the domestication of wheat and barley in the Levant had not yet been reached, and in turn that process was similarly interrupted by the cold of the Younger Dryas. The two developments, in the Levant and in China, were wholly independent of one another, and went on simultaneously, though the Levantine process seems to have actually started first, and was successful well before the best processes of rice cultivation in China were developed.

The domestication of animals was also slower and different in China than in the Levant. The first animal to be domesticated in China seems to have been the pig, almost as soon as the success with rice, and the process seems to parallel that of rice; in the Diaotonghuan Cave the early layers contained not just wild rice remains, but those of wild pig as well. The later layers, with the suggestion of domesticated rice, also contained the bones of probably domesticated pigs; they were soon joined by chickens. The choice of animals was clearly different in China and in the Levant – further evidence of the independence of the process in both regions.

Maize

There are two other areas where plant domestication took place. One, Central America, was clearly wholly independent of the rest of the world, even though the process began several millennia after the parallel processes had been successful in the Levant and in China. The earliest signs of agriculture, archaeologically, are in the Panama and Mexican regions, perhaps before 6,000 years ago (a good two millennia after success had been achieved in China and the Levant). A considerable variety of plants was eventually cultivated in pre-Columbian America, but the most important in the end was maize. This was a plant which is very difficult to domesticate, but one which was very productive in the end, once it had been bred up to provide a substantial food quantity. It seems probable that the early, small, maize plants were deliberately bred to increase their productivity, implying a preceding understanding of what was required.

Other plants domesticated in America were various squashes such as tomatoes, beans, manioc, potatoes, and the sweet potato. The number of successful domestications is large possibly because of the absence of a grass-type staple like wheat or rice, for maize is less productive than these crops. It also seems that some of these crops were domesticated several times, and so in several different places. This suggests that it was the idea of domestication and cultivation which became widespread, and that only the difficulty of finding a suitable plant to rely on, and operate on, delayed matters. Maize eventually won the race.

African Domestications

It is also possible that plant domestication took place independently in sub-Saharan Africa, but the current evidence is that agriculture did not begin to be practiced anywhere south of the Sahara before about 3500 years ago. The crops developed – not necessarily all at an early period – were certainly different from those in the rest of the world: pearl millet and sorghum particularly, also African rice, yams, and finger millet. All of these were native to different parts of Africa, so at least the idea of domestication had reached Africans by that time, and this implies that when the idea arrived it stimulated Africans to search out likely plants

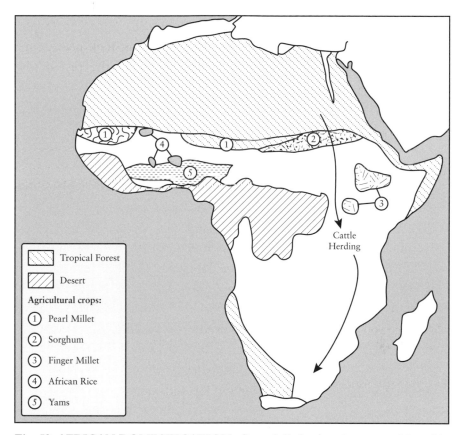

Fig. 59. AFRICAN DOMESTICATION. Several distinctive plants are cultivated in Africa, a pattern which strongly suggests that the idea of farming arrived, and local plants were then selected for domestication. At the same time, and surely not coincidentally, the practice of cattle farming moved south from the Sahara and the Nile Valley into East Africa and then on southwards.

for domestication; probably there were failures as well as successes. It was almost always possible to travel across the Sahara and along the Nile Valley, or east-west through the savannah south of the Sahara, and the notion of domesticating plants could thus have arrived from various Mediterranean regions, Egypt, North Africa, Spain.

It is also quite possible that the Africans domesticated their crops independently. The long delay in doing so suggests that the need for domestication was not perceived for a long time, that, in other words, Africans could meet their food needs by other means. In East Africa the domesticated cattle, whose herding spread south from the Sahara, would be one of these; foraging in the forests was also a successful strategy. The likelihood is that domestication of plants happened because a renewed dry phase in the Sahara was putting increased pressure on the resources available to the sub-Saharan hunters and gatherers. This is, of course, just the theory of the origin of the preceding domestication of cattle in the Sahara (Chapter 6). It is noticeable in this connection that one of the oldest sites where agriculture was practised in Africa is in West Africa, in Mauretania, an area where the desert would quickly encroach onto the savannah – the same sort of marginal area which featured in the domestication of wheat and barley in the Levant. (Mauretania is close to Morocco, which could have been a source of the idea of domestication.) The expansion of the desert was certainly driving the cattle herders southwards in the direction of the Great Lakes region in East Africa at that time. It would seem, therefore, that the domestication process was here a response to the end of the Ice Age, just as it was elsewhere, though the pressure had taken several more millennia to become effective. (As it had, of course, in America.) But then it took several millennia for the response to the invention of agriculture to have its full effect in both the Levant and China (Fig. 59).

New Guinea

There is one further region where plant domestication took place: New Guinea. It happened there despite the fact that the island is close to the equator, and despite the further fact that it was in the Highlands of the island, well away from the rising sea level, that the work took place. It was certainly a direct result of the effect on the island of the change of climate at the end of the Ice Age, and equally certainly a development wholly independent of any other region. In fact, it will be seen that almost exactly similar conditions operated here as elsewhere; the main difference was that the climatic and vegetation change is a vertical event rather than a horizontal. But this enforced another difference, which pertained to the plants involved.

The New Guinea Highlands, whose peaks range up to 5,000 metres above sea level, and large parts of which are over 2,000 metres, were ice-covered during the Last Glacial Maximum (Fig. 60). One effect of this was to drive down the altitude at which trees could grow. The forests of

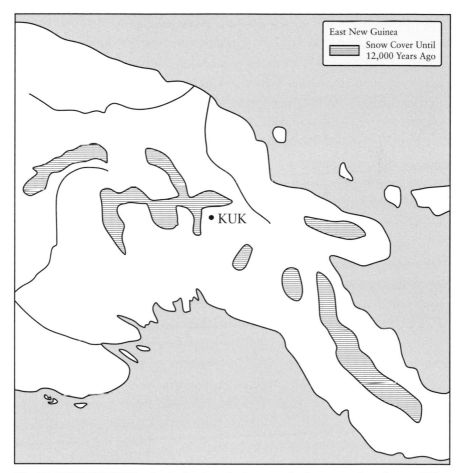

Fig. 60. NEW GUINEA. In the Ice Age the ice cover in the mountains of central New Guinea was extensive and the forest cover correspondingly lower. Kuk, where evidence has been found for early agriculture, is situated on the margins of the possible cultivation of the New Guinea plants in land which was at that time (about 10,000 years ago) open forest.

the island contained a number of plants and fruits which had long been collected by the inhabitants, whose ancestors had, of course, been in the island for several tens of thousands of years already (see the Prologue). They were surely very familiar by now with the forest and its products. The effect of the cold of the Last Glacial Maximum was to increase the ice caps on the central mountains, and so to push the successive bands of vegetation down to lower levels on the slopes of the mountains. So the snow line was pushed down to about 2,200 metres above sea level from the 3,900 metres it had been at beforehand (and is now). Below the ice was a sub-Alpine land of open ground, with low-growing grasses and herbs and flowers. This is a sort of high-level steppe, a land where grazing marsupials throve. Below this again were the successive types of trees down to the tropical forest, which occupied the lowest slopes close to the sea level: this also implies a rapid increase in the normal temperature, from the frozen ice caps to the seashore, and over a relatively short distance. For hunters and gatherers all these territories, except the ice and snow regions, were useful and productive, the forest for its fruits and nuts and roots, the more open lands for pursuing and hunting the indigenous animals.

As the general temperature rose with the ending of the Last Glacial Maximum, so the boundaries of these vegetation bands moved up the mountain slopes, but at different rates. The snow retreated more rapidly than the bands of forest advanced, and so the intervening bands of vegetation – the sub-Alpine open land and the non-tropical forest – temporarily expanded. For the human inhabitants this meant that a greater area of land was available for them to exploit than before. The expansion of the open sub-Alpine lands increased the area over which they could hunt; the expansion up the mountains of the open forests increased the area over which they could gather forest products. The human population would seem to have grown as the resources they used expanded.

One human reaction seems to have been to develop the use of fire, perhaps to clear woodland and forest into which grass could spread and so provide larger areas where herbivores could graze; this would force the growth of fresh grass for grazing animals, or perhaps drive out huntable animals into the open areas, where they could be the more easily killed. In all cases, of course, the object was to manage the open woodland in order to increase the number of animals available for the hunt. But it also opened up the possibility of using the same areas for growing crops when the techniques of farming became invented and known.

Within the forest the hunters and gatherers had long exploited all the available foods. New Guinea had such useful products as taro and yams – both root crops – *Australimusa* bananas (a type of banana with vertical stalks), sugar cane, and other useful plants. Taro is particularly intriguing since it needs extensive preparation to render it suitable for humans to eat, including scraping, soaking, chopping, and cooking, in order to remove unpleasant or poisonous elements. That this sequence had been achieved long before the Last Glacial Maximum is shown by the discovery of taro remains at very early sites in the islands of the Bismarck Archipelago. This, of course, does not mean that the crop was farmed, only that it was available to be gathered in the wild. The same may be said of other forest products, but the sheer persistence and experimentation which was involved in working out how to use taro as a food is a testimony to human ingenuity and determination – but we do not know how many died of wrongly prepared taro in the search. It is also, of course, a mark of the desperation of the experimenters; one would not go to all that trouble if an easier crop was available.

As the range over which the land became forested grew and forests gradually expanded up the hillsides and valleys of New Guinea, so the altitude also rose at which tropical forest vegetation like bananas and other local crops could thrive. It seems that it was at that point that men resorted to deliberately preparing the ground in which these plants will grow. It was therefore most likely a clear development of earlier processes of manipulating woodland and grazing land to increase their food supply. At Kuk in New Guinea excavations have found that drainage channels, with overflows and control pools, were excavated in a swampy area before 9,000 years ago. These channels were later covered by an increase in the deposition of clay from further up the mountainside at about that time – perhaps caused by the increased flow of meltwater – so the channels are obviously earlier than this well-dated deposit (Fig. 61).

The object of the channels and the pools was to control the drainage from the swamp, and so enable their chosen crop to be planted and to grow, and presumably be tended and weeded. The same measures were instituted in the same area on several occasions later, and the dating is firm. What the crop was in the earliest phase is, however, not clear. Any of the local domesticated plants – taro, yam, sugar cane, bananas, pandanus – could have been grown, or perhaps it was very likely a mixture of several. The swamp was at such an altitude that these plants could not have been grown in that area during the Last Glacial Maximum, but they could have been grown there once the warmer following period had arrived.

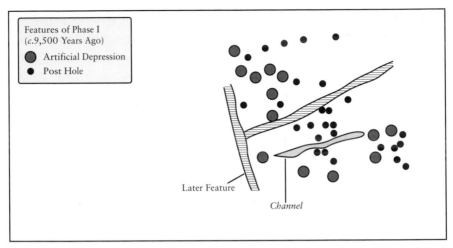

Fig. 61. KUK. The evidence that Kuk consists of an artificial 'runnel' which was presumably for drainage purposes, several stake holes, which are clearly an artificial feature, and a number of artificial hollows which represent the places prepared for planting the crop. The diagram illustrates the remains of phase 1 (about 10,000 years ago); two later phases were also found, showing much the same pattern remains.

It would seem, therefore, that the new farmers were taking advantage of the change in climate and so of vegetation to institute their own planting programme. This would secure for them a dependable supply of food to supplement the product of their hunting. This in fact may be a type of foraging, where cultivation was being used to expand the possibilities of collecting foods, but the better term for it is horticulture. The obvious next step would be to establish a permanent residence nearby, and to farm other crops. The evidence for all this is still fairly tenuous, based as it is essentially on one excavation, but it does imply that the domestication of food plants was taking place in New Guinea in the period immediately following the end of the Ice Age, just as it was in China and the Levant.

This is clearly a completely separate, indeed a wholly different, process than any of the other domestications. The plants involved – taro, yams, *Australimusa* bananas, pandanus, and so on – are unique to the area, and were planted individually rather than the seeds being broadcast, as is the case with the grasses. The methods of cultivation were therefore essentially horticultural rather than agricultural. The plants had to be placed carefully and singly by hand, watered, and the weeds removed. The principal tool in use was a digging stick, used to make the hole for the plant and later employed as a hoe. (This, of course, became the preferred method of the cultivation of rice, but there does not seem to have been any connection between China and the much earlier horticulturalists of New Guinea.) These methods were not employed elsewhere in any other regions except to cultivate relatively small areas of vegetables; it did become the practice to flood rice paddies, though this is not really the same technique. This development in New Guinea was a wholly independent matter of increasing and managing the local food supply, and, while the details of the processes and methods are not yet wholly clear, the timing of the process indicates that it was a human response, once again, to the climatic changes brought about at the end of the Ice Age.

One other aspect of New Guinean agriculture was that it did not spread to other regions (unless taro growing did spread into the islands to the east, or perhaps from the east into the island, though the plant seems to be indigenous to both New Guinea and the islands). Rice and wheat agricultures, however, both proved to be exportable to many other areas. From the Levant the farming of wheat, barley, and domesticated animals spread east and west into Iran and Europe; from central China rice agriculture spread south into southern China and south-east Asia and Indonesia, into Japan, and eventually to India. These responses to the global warming of the ending of the Ice Age proved to be the most successful of all. But New Guinea's practices, perhaps because the methods were time-consuming and laborious, did not travel; it is also the case that New Guinea, particularly the interior territory, was unusually isolated from the rest of the world, and indeed, its many tribes and clans tended to be isolated from each other as well. However, it is apparent that this isolation was not complete, because some of the foods, the bananas in particular, but also taro, did spread to other areas. Like American pre-Columbian domestication, in New Guinea the process domesticated a considerable number of food producing plants, so that, having succeeded with one plant, others were experimented with; it was clearly a successful development.

Agriculture's Empire

The invention of agriculture, and in so many places independently – the Levant, China, New Guinea, Central America, though in West Africa it must be assumed that the idea arrived and was locally adopted – proved to be the most productive of all the Mesolithic period adaptations to the changing global climate. It proved to be a method of food production which could be extended to other lands without too much difficulty. Already in the earliest years the cultivation of barley and rye had extended from Syria into Anatolia and Iran; soon it would move into Babylonia and Egypt, where the reliable water supply from the great rivers combined with the rich soils deposited annually by the rivers' floods would increase the plants' productive capacity many times over. In the same way rice farming could be spread throughout South and Southeast Asia, and maize could be grown in much of South and North America, both in the dry lands of New Mexico, and the damper, cooler mountains of Peru and Mexico; South America's potatoes were also a plant capable of spreading widely. Many of these responses extended into regions unsuitable for cultivating the original domesticated plants but where they could be bred to wider tolerances, and also where other plants could be domesticated – squashes and potatoes in America, millet in north China, vegetables everywhere.

This response proved to be the winning characteristic. The other human responses to the new climate, while successful for small communities, were, so to say, non-exportable. Foraging was always a local activity, and could be done on any coast with a supply of shellfish, but was limited in many areas; only in Japan does it seem to have been successfully systematized, as a result of the unusually compact geography of the localities. But molluscs are a less nourishing food than most, and foraging for such shellfish can only support a fairly small population. More promising was the domestication of cattle, but this proved to be most useful only when combined with the new agriculture, as had happened with the more or less simultaneous domestication of plant crops and sheep/goats in the Levant. However, a pastoral life usually required constant movement, and can be seen as a logical development from a hunting lifestyle, though it did take some imagination to shift from hunting to herding. Agriculture, the growing of wheat and rice and so on, necessitated the same sedentary lifestyle as foraging, and produced the same large and reliable quantities of food as pastoralism. It was the combination of these two characteristics which allow the development first of villages, and then in especially productive regions, of towns and cities, and the increase in wealth and resources and skills which this implies.

Above all, of all these new ways of gaining a living, it was agriculture which permitted a growth in the human population. The availability of a reliable food supply meant that life became easier and less stressful, and the necessity of staying put allowed the better survival of infant children, and a longer and healthier life for adults, also the accumulation of possessions, including permanent housing; this in turn reduced their exposure to inclement weather and so assisted in improving human health, both of the community and of the individual. The sedentary life was less physically demanding also. With an assured food supply, an easier life, more comfort, it is no surprise that agriculture was welcomed wherever it spread.

Perhaps the most convincing element in this improvement is the comparison of the expectation of life for the Ice Age and Mesolithic hunters, most of whom lived no more than thirty years, with the much longer lives expected of their later descendants, working in agriculture, to the reported expectation of seventy years, suggested in the Christian Old Testament, the product of an agricultural peasant society.

Further Reading

Peter Bellwood, *The First Farmers*, Cambridge 2005, is a reasonably up-to-date and accessible account of the invention of farming, with a huge bibliography, and covering all the world. For the Levant, an older account going into great detail, is James Mellaart, *The Neolithic of the Near East*, London 1975, also with a good bibliography. For the Natufians, see O. Bar-Yosef and F.R. Valla (eds), *The Natufian Culture in the Levant*, Ann Arbor, MI, 1991; for Abu Hureyra, A.M.T. Moore, *Village on the Euphrates; from Foraging to Farming at Abu Hureyra*, New York 2000.

For the history of rice in China, X.J. Zhao, 'The Middle Yangtze region in China is one place where rice was domesticated: phytolith evidence from the Diaotonghuan Cave, northern Jianxi', *Antiquity* 72, 1998, 885–897 and H.Y. Lu *et al*, 'Rice Domestication and Climatic Change: phytolith evidence from East China', *Boreas*, 31, 2002, 378-385.

For New Guinea, see three articles in the *Antiquity* 'Transitions' supplement: by Geoff Hope and Jack Golson (Golson was the excavator of Kuk), D.E. Yen, and David Harris, plus T. Fenman *et al*, 'New Evidence and Revised Interpretations of Early Agriculture in Highland New Guinea', *Antiquity* 78, 2004, 839–857, and T.F. Denham *et al*., 'Origins of Agriculture at Kuk Swamp in the Highlands of New Guinea', *Science* 301, 2003, 189–193.

Gallery VII

The Invention of Agriculture

The invention of arable agriculture took place more than once, but at first only in the Levant. The earliest case is at Ohalo, beside the Sea of Galilee (VII.1 and 2).

VII.1 Reconstruction of a hut at Ohalo.

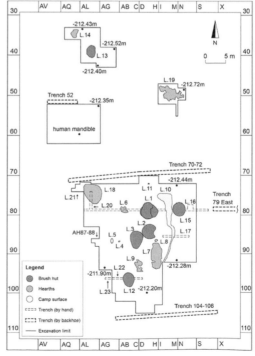

VII.2 Plan of Ohalo excavation.

The earliest success was also in Palestine, where the excavation at Natuf revealed a society clearly depending in large measure on farmed foods, as indicated by the pestle and mortar, used for grinding the product into flour (VII.3). It was a widespread culture, covering all Palestine and into Syria to the north.

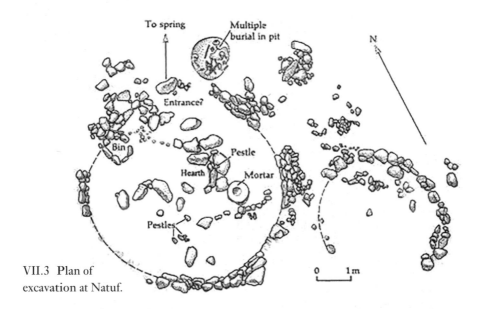

VII.3 Plan of excavation at Natuf.

VII.4 Excavations at Jericho (not far from both Ohalo and Natuf) aimed at finding evidence relating to the bible, actually discovered a 'city' dating back much farther, to about 9,000 years ago.

VII.5 Gobekli, Pillar. The discovery of Gobekli is another surprising development, revealing an elaborate series of round structures containing t-shaped pillars, many of which have low relief carvings on them. The place is one of half-a-dozen other settlements displaying the same type of structures, whose precise purpose is unknown, though religion has been suggested. Yet there is no real indication of worship or any other sign of religious practice.

VII.6 Gobekli, 'circular structure'.

VII.7 Gobekli, partly
excavated T-shaped pillar.

Conclusion

The onset of the global warming as the Ice Age ended eventually brought about more congenial conditions of life to much of the world, but the change was slow and erratic, with the warming interrupted at times by renewed cold periods, particularly the Younger Dryas period. The change from the stable conditions of the Last Glacial Maximum to the similarly stable conditions of the so-called Climatic Optimum of the Holocene took perhaps five millennia, and even after that there were relatively minor fluctuations, such as the warm period of the Bronze Age and later 'Little Ice Ages'. It was during this period of the major changes, the Mesolithic, that the ingenuity of mankind in all parts of the world, from Australia to Norway, was put to the test. They were busy devising ways of coping with the change; these are the efforts discussed in the previous chapters – and, of course, the 'change' varied from place to place, as did the results of the efforts made.

I make this point because all too often the archaeological evidence is considered area by area and in chronological segments, whereas here I have been looking at the whole world over a relatively limited period of time (and therefore in a fairly summary fashion). It is salutary to realise that, when farming was being invented in the Levant, an ice sheet still covered much of the Scandinavian Peninsula, the North Sea was dry land, and that when horticulture was being developed in New Guinea the peopling of Scotland had hardly begun. What does emerge is that the ending of the Ice Age brought decisive changes in all parts of the world – the disappearance of the ice, the expansion of forests and of deserts, by the rising of the sea level and flooding of lowlands, and the changing of the weather patterns. So the invention of farming in the Levant and China and of horticulture in New Guinea, the development of successful foraging lifestyles in Denmark and Japan and the Scottish islands and Peru (and other areas as well), the domestication of cattle and sheep and goats and pigs and hens in the Sahara and the Levant and China, and the settlement of the Fish-eaters on and near the shores of the Gulf, all took place within that period of five millennia of unstable climate, or shortly after, and such developing alterations in human lives began to happen when the effects of climatic change struck that particular area.

I have several times made the point that the changes in human life which we can see taking place in that long period were driven in all cases by the wish of people to change as little as possible – the conservative reaction, which seems to be a universal impulse. The hunters and gatherers of the Palaeolithic/Ice Age wished to continue to be hunters and gatherers as the ice retreated. So the mammoth hunters of Ice Age Europe, some of them at least, simply moved to new hunting grounds with their prey, and there continued living and working in the same way. The hunters living in Ice Age Beringia found their way south into the rest of America and continued their hunting life in a region of much more varied landscape and with a much wider variety of animals to prey upon – but they lived in the same way as before, if rather more comfortably and wealthily; those who stayed in the shrunken Beringia also changed as little as possible, one group taking to the ice to continue their hunting lifestyle, another set turning to the sea, but still hunting, though their prey was in the sea. In Western Europe people living in Doggerland shifted their homes to Norway in order to stay in the same sort of climate, cold and wet though it was, though here they were compelled to rely more on fishing for a livelihood, as were the Aleut and those who moved to occupy the American north-west. Even so, this was probably a fairly minor change to their way of life for them, since their lives in Doggerland had latterly been dominated by the encroaching sea and by the great rivers.

The wish not to change one's life very much is an intelligent response to outside pressures, especially if the prospect is unpleasant. Life for hunter-gatherer groups was hard and fairly short. They relied on gathering food and hunting animals for immediate consumption, though the cold of the Ice Age might have allowed them to use the frozen winter for some storage. But this was a life lived on the constant edge of hunger. It would never take much – a flood, a drought, an exceptionally long winter, an accident – to drive these hunters into starvation and death. Few skeletons from the Ice Age show a longer life than 30 or 35 years. In this difficult and dangerous life, therefore, any change had to be studied with great care before valuable energy was to be spent on the change, so that it was accepted and incorporated in the life of any hunting band. Conservatism was not merely a preference, it was a life-saver.

On the other hand, these conservative families in attempting to conserve their lifestyle in the new conditions, were in many cases compelled in the end to adapt and change because the new conditions were intolerable; conservatism can also be a death-trap if changes are too fanatically resisted. So the Doggerlanders had not merely to move to Norway, but they had to turn more to the sea for sustenance than to the land, because whatever animals they had hunted in Doggerland probably did not exist in their new homeland. Others affected by the changes were compelled to change more drastically, especially if they stayed in their former homes and faced the change. So the Fish-eaters stayed in the Gulf area but were driven to exist almost entirely on fish. In Japan the Palaeolithic population continued to pursue both hunting and gathering, but for the groups who settled into relatively small areas, thereby becoming foragers over much more restricted regions, found their diets favouring vegetables and fruit and fish over meat. In Denmark the proto-Danes were similarly settled into one relatively small region, and concentrated largely on fishing whereas their ancestors undoubtedly ate mainly meat. These foragers therefore shifted their lives from a wide-ranging and mobile hunting existence to where they were settled in a relatively restricted area, and they had to shift their diets away from a heavy reliance on meat to a more varied one including more fish – or almost entirely of fish, in the case of the Gulf Fish-eaters and the Norwegians – and more vegetable foods and fruit. This constituted a major change in their lives, as great a change as perhaps they could have possibly imagined. It was also, as it happens, a much more balanced and healthy diet, a case of an inadvertent improvement brought about by a reluctant change.

The same may be said of those groups which eventually developed into farmers. The people of Ohalo, the Natufians, and the Geometric Kebaran people in the Levant were all originally hunters and gatherers, and did not abandon that life, but their lives shifted gradually into foraging, and their gathering activities came to concentrate on harvesting the wild grasses, as they came to rely less on hunting. This sedentary activity came to dominate their lives as their numbers grew and as they moved less. Once a productive area had been located they would need to guard it, from both grazing animals and other humans – though they will have found that grazing animals who approached would be relatively easy to hunt, another inadvertent advantage of the change. The need to protect their lands and crops restricted their movements in the same way as the fishermen of Denmark were pinned down to a few places by their fish traps and productive areas. For ten millennia the Levant grass harvesters were just as much foragers as the Japanese of the Jomon period, and, as it happened, both of these foraging periods lasted about the same length of time. In China it seems probable that a similar foraging period occurred, though it lasted for less time, since the emergence of farming occurred soon after the end of the Ice Age, as it did in the Levant. Successful rice harvesting requires the use of pots for cooking, and so one may perhaps assume that pottery making and rice harvesting went together, and the existence of pottery probably speeded the adoption of farming. In China the original foraging developed into farming, whereas in Japan the foraging remained sufficient, perhaps until the population grew to such numbers that a more adequate, greater, and more dependable food supply was needed – hence the adoption of rice farming.

These non-farming expedients – moving to new hunting grounds, foraging, fishing – were successful in their purpose, which was to ensure a sufficient supply of food for the families taking up the new life. A lifestyle which lasted ten millennia, as the foraging did in more than one place, was surely a successful lifestyle; the new Norwegians and the Fish-eaters established societies that lasted almost as long, indeed in a way their fishing activities might be considered a form of foraging, though hunting would probably be closer; the Inuit and Aleut did continue to be hunters but they developed a version of hunting which was stable and productive, and lasted for millennia; the mammoth hunters who moved with the mammoths into Siberia lasted as long as their prey existed, which was probably another five millennia after the end of the Last Glacial Maximum – this was also a stable adaptation to the new conditions.

As responses to the end of the Ice Age these all indicate that people were adaptable; even when attempting to remain the same, they were capable of adapting to new conditions. Some groups could, in fact, choose from a variety of possible futures – continue hunting, become foragers, or turn into fishermen – and it is very likely that many of the changes were made consciously, with some idea of the consequences, by copying from others who had already changed, but certainly by thinking through the consequences with care and in advance. Moving north over a long distance in order to continue hunting the same animals was clearly a deliberate decision; the move by the Doggerlanders to Norway was something which had to be planned and organised for it to be successful – and it clearly was successful.

At the same time, people everywhere were still very much subject to the vagaries of the erratic climate. The repeated failure of the resort to grass seed harvesting in the Levant, in some cases clearly as a result of the onset of the Younger Dryas cold, is one indication of this, as is the long gap in the use of rice at the Diaotonghuan Cave. But it is worth noting that the experimentation in grass foraging went on over a long period of time in both lands; the idea was clearly present among the hunters and gatherers of the Levant all the time from the Ohalo experiment onwards, and it may be assumed that there were other attempts at solving the problems involved – Abu Hureyra was clearly one of them, a major experiment which failed. Similarly, in America the development of the cultivation of maize went on over a long period until the plant was bred up to provide a substantial food supply. The record at Diaotonghuan suggests that it took several millennia in China to transit from foraging/ hunting to farming. No doubt it was the same in New Guinea. It seems clear that it was widely appreciated that agriculture was the way to go, and that repeated attempts and experiments would eventually bring success. A considerable knowledge of the plants and their lives must have been accumulated. It is often remarked that the accuracy of the cave paintings of animals was presumably the result of the hunters' intimate knowledge of their prey; those who developed their chosen plants into farmable crops seem to have displayed a similar knowledge – these were probably the women, whose task seems to have been the gathering of plants, while the men went hunting.

The result of all these changes, and of the responses of different groups to the pressures to which they were subjected, was that new and varied societies developed in many parts of the world. The divergence between the societies, into hunters, pastoralists, farmers, foragers, fishermen, was something new in human history. Human development had gone in parallel in all parts of the world, with the universal pursuit of life as hunter–gatherers. Towards the end of the Palaeolithic, for example, the use of microliths, or microblades, had developed in many different areas, more or less at the same time. They can be seen in Europe, in India, in Beringia, in Japan, and in Australia. It is not reasonable to assume that these areas were in communication with each other so that the ideas of new tools and so on could be passed along – certainly Australia and Beringia were out of touch with all the rest. The simultaneity is therefore presumably a normal human reaction to similar events and pressures – the dry cold of the Last Glacial Maximum. Other similar and more or less simultaneous developments and inventions seem to have included dugout canoes, bows and arrows, and the early domestication of the dog (which appears in Japan and Europe at about the same time). The invention of pottery may seem to have been lagging in the Levant behind the developments in the Far East, but in fact the time-lag was about four millennia, which may well be the same sort of time difference in the adoption of these other inventions and developments – and the invention of pottery in the Far East did take place, like all the others mentioned, during the Last Glacial Maximum.

In the same way, there are examples of similar developments or changes in human behaviour in different areas. The respectful burial of the dead appears to be a basic human practice, and the use of red ochre scattered over the body is widespread, and is known in Australia, in America, and in Europe. This also involved in many cases the practice of leaving gifts with the dead, and eventually into the practice of concentrating the burials into cemeteries. These changes and practices may be connected with the increase in the population and of larger groups living a sedentary life together, which was the result of the increased skill at food gathering, and which was in part the result of these inventions. Foraging in its various manifestations enforced living in one place and promoted the growth of larger societal groups, and was one of the developments in the warmer Mesolithic. The concentration of the dead into cemeteries mirrored, and was the result of, the concentration of the living into sedentary settlements.

But what is most telling is that the developments happen everywhere, wherever people became more or less sedentary – in Mesolithic Europe, in the Levant, in Jomon Japan, in the Gulf, in China, in India, in Australia. This similarity of response occurred first at the end of the Ice Age, as part of the impulse to conserve as much of the old Palaeolithic life as possible in the face of a threat to it. But the responses soon diverged, into developing a foraging lifestyle, into moving with the ice, into moving away from the ice, into taking to the sea. It was the adoption of a sedentary existence, however, which was the most profound change. It was this which was necessary for a successful foraging life, for that practice required an intimate knowledge of the food resources of a limited territory, from which regular harvests of food – shellfish, fish, nuts, fruit, animals, and other crops – could be taken without depleting any of these resources.

Sedentarism was thus a successful ploy to cope with the problem of the new climate, but once adopted, the new lifestyle had further consequences. Foraging proved to be conducive to an increase in the population, since the life was less rough and wearing and stressful and accident-prone than the constant movement which hunting required, while food was relatively more plentiful and reliable, and healthier. This was especially noticeable in the Levant, where the Natufian period saw a substantial increase in the numbers of villages practicing the Natufian way of life, which was sedentary farming, but which also included a hunting element, though it seems that Japan, Denmark, northern India, Oronsay, Peru, and China all saw the same phenomenon. This greater population in turn was the foundation for the next development, the definitive invention of farming in the Levant, one of whose drives was the need to feed the larger population evident in the increase in the number of the Natufian villages. The same happened in China, probably from the same cause, which was a combination of the greater population (and hence a certain pressure on resources) and an adverse change in the climate. These groups were larger both in number and in population than the single families of hunters. This was the new society.

The climate was, of course, a very controlling factor in all the various endeavours undertaken by these different groups of people. It was the change in the climate which propelled them into the changes they had also made in their lives, and it was the subsequent erratic climatic changes, in particular the recession of the Younger Dryas, which interrupted or side-tracked those endeavours. But it was not the only thing involved. Had the climate been wholly paramount as a cause the reactions of the people would have been similar everywhere; instead their reactions diverged into the different lifestyles, which will have begun from deliberate decisions by the people themselves. That is, the people were certainly subject to powerful climatic influences, but they were developing and organising their own lives as well, and their responses were the result of individual or group choices and decisions. The climatic change was the impulse, but the people made their own choices on how to cope with the change.

This is the real change, for the climatic changes compelled individual groups, and indeed perhaps individual people, into adopting a new style of life, and to do this consciously; it also made it possible to choose that new life from the options which they knew about or could imagine. And many of them seem to have succeeded. All the new lifestyles which emerged after the Ice Age remained successful for several millennia. Those who followed the mammoths to the north continued to hunt them and the other animals of the cold lands, and went on doing so when the mammoths themselves were extinct; their descendants are living in Siberia still, and hunting (having, no doubt, adapted to the new condition of the absence of their primary prey). The foragers in various places lasted for a long time also, the Japanese of the Jomon period establishing a way of life which lasted for ten millennia. The proto-Danes did not last so long, being overtaken by agriculture well before the same happened in Japan, but they certainly preserved their foraging life for perhaps four millennia, and the foragers who used Oronsay as their chief foraging ground seem to have done so for about the same length of time. The cattle herders of the Sahara were, of course, rather more directly influenced by the changes in the climate than most groups, but they were able, by the very mobility of their animals, to move off to other lands which remained moist when the desert dried out, taking with them a mobile food resource. The Fish-eaters of the Gulf originated perhaps 9,000 or 8,000 years ago, and there were still representatives of their life style living in Gedrosia, if not elsewhere, for Nearchos to attack in the late last millennium BC.

Two of the solutions to the difficulty of finding sustenance, however, had greater success than mere longevity (though any method which lasted even one millennium – thirty to forty human generations – let alone for four or ten millennia, must be counted successful). Some of the cattle and sheep herders of the Sahara migrated south as the desert advanced, this led them either into the savannah of West Africa or towards the Great Lakes region of East Africa. Some became the ancestors of such groups as the Fulani who have roamed the savannah in West Africa ever since; those who moved towards the Lakes region eventually found a way through, between the rainforest and the tsetse fly of the Congo basin on the west and the Indian Ocean on the east, and spread their way of life all along and across southern Africa, moving south in a fairly leisurely way until they reached South Africa about the end of the last millennium BC, becoming ancestor of the cattle-keeping nations such as the Zulu and the Xhosa.

The agriculturalists of the Levant and China, however, were the most successful of all in their response to the problem. Once the techniques of farming had been worked out and were understood, they proved to be eminently exportable. The techniques were taken either personally into new lands by migrants, or those techniques were adopted by others either to copy the originals or to develop and domesticate other crops. Having taken several millennia to develop in its homeland in the Levant, the farming of wheat and barley and so on proved to be adaptable to colder and wetter climates, and beyond those, rye could be grown. The new technique reached the Balkans and the Indus Valley about 8500 years ago, but was delayed in reaching Egypt by the barrier of the Sinai desert. By 6,000 years ago there were farmers in the British Isles and Spain, and in the Nile Valley and Iran as well – and in all these places they proved capable of building even greater monuments than Gobekli – the Pyramids, Stonehenge, the Malta 'temples', and so on. India beyond the Indus valley proved to be less susceptible than Europe; there the differences in climate and the seasons was a major barrier, as it was in Egypt, but farming societies, combining tending crops with herding cattle, reached into the south of the sub-continent and into the Ganges valley by about 5,000 years ago.

The Chinese agricultural system, based on rice, spread rather more slowly, for it was a more difficult technique, with its requirement for irrigation in its developed form, and hence it needed more preparation;

it was a more complicated matter than the cultivation of the western grasses. However, most of China south of the Huai River was producing rice by farming by about 5,000 years ago, and by that time the technique of rice cultivation was also being used in the Burmese delta region as well. This agricultural system reached into India about the same time, moving into the Ganges River area, where it met the wheat-and-cattle economy from the west. The most productive areas in both wheat and rice cultivation turned out to be Egypt, Babylonia, and the Ganges valley for wheat, and the Yangzi region and the Burmese delta, where reliable and copious supplies of water were available, for rice. This is odd, since it was in a semi-desert region that the technique of wheat farming, and in a region which was on the edge of rice's range, that the farming methods were first developed and were proved to be successful. Ultimately, rice growing conquered much of Southeast Asia and the Indonesian archipelago, just as wheat and barley conquered Europe and western Asia, but rice took longer in its march of conquest than wheat, no doubt due to difficulties of establishing and learning the cultivation methods.

The New Guinea system of horticulture proved to be similarly exportable, as were the island's domesticated crops. Taro especially became a staple crop of the Pacific Islanders when they moved out from the eastern end of the Indonesian islands, and yams and bananas spread well also; the methods of New Guineans were in use eventually in the Hawaiian Islands.

The success of the agricultural systems can be measured partly by their successful transplantation into other areas which had somewhat different climates and peoples, but also by the fact that they made it possible to support constantly increasing numbers of people. In the process of expansion, of course, farming also overwhelmed its alternatives: the foragers of the various places where that life had been adopted were unable to compete, and sooner or later many of them succumbed to the wealth-creating reliability of agriculture. The cattle herders of Africa, however, proved just as successful in their expansion, and another version of their lives, the horse herders of the Eurasian steppe, were similarly adept at spreading their lifestyle. Even in these cases, however, the farmers eventually won.

The success of agriculture has sustained the expansion of wealth and population for ten millennia in the Levant, and for lesser times in areas where it colonised later. It has, of course, now reached a clear limit. The success of the agricultural regime in supporting increasing numbers of people has probably just about reached as far as it can go, and by doing so it has raised a new major problem, for one of the causes, perhaps the major cause, of the new episode of global warming which is affecting the world now is overpopulation. We have come full circle.

Lessons for our Present Predicament

The avowed purpose of this book has been to examine some of the human responses to the global warming episode at the end of the Last Glacial Maximum, with the aim of seeing if any indications can be derived from those ancient responses to assist in our own response to the present episode of warming which appears to be overtaking us now. I have used this term 'lessons' as the heading for this section only after failing to find any less didactic alternative, for historians are always wary of such a concept as deriving lessons for the present from the past. It is far too easy – and common – for those with agendas of their own to seize on occasional or isolated historical 'facts' for their own purposes – the 'history tells us' syndrome. The distortions and selectivity used by Nazis and Fascists and Communists are themselves the real 'lessons' of the dangers of doing so, though that hasn't stopped other politicians from casually using it. They were all put in their place by one of the heroes of technological history, Henry Ford, when he is said to have remarked that 'history is bunk'. It is not the purpose or task of historians to use the past to prescribe for the future, but it seems that in this case, since the human population of the world has gone through one episode of global warming, it is worth seeing how they did it, which is the previous content of this book, and then to see if any general – not specific – 'lessons' can be drawn. (Actually we have gone through more than one, as the first Australians could attest.)

It is worth noting that those who are most shrill in the argument that global warming is happening and is a disaster in the making are themselves very liable to such distortions and selectivity. The temperature changes they suppose will happen are nowhere near as great as those which took place at the end of the Ice Age, or in earlier episodes; their historical basis of comparison therefore normally ignores that event and concentrates on the last hundred years or so, when, they claim, the records are sufficient and accurate. They usually illustrate it with a graph of predictions which exaggerates the vertical scale and so over-emphasises the size of the supposed change. But anyone who studies history must be just as wary of predictions as of 'lessons', and must avoid such facile short-term comparisons. The one certainty about predictions is that they are always wrong, and if any do turn out to be correct it is on the principle of the stopped clock – it will be right twice every day. One of the additional purposes of this book is to assist in providing a longer perspective to the discussion.

There are thus no precise 'lessons' to be learned directly from our knowledge of how people coped with the changes at the end of the Ice Age. We cannot look back and pick out a response which was successful then, and apply it to our condition. After all, the most successful response was the invention of agriculture, which can hardly be repeated. The conditions of life in the world have changed sufficiently since the crisis of the Mesolithic that it is not reasonable to equate the reactions of men emerging from the Ice Age with those living in the Industrial Age. But it seems clear that mankind itself has not changed in any serious respect, despite the economic and technical changes he has achieved. The creature who emerged in the Ice Age as a hunter and then became variously a forager, a fisherman, a farmer, a herder, is recognisably the same adaptable creature as the one who now works in factories and offices and hospitals and schools, who lives in houses, and who travels and moves, and who searches for a means of sustenance. All this is clearly an inheritance from the cave-dwelling, nomadic hunters of the Ice Age and before. It is also obviously a convincing inheritance to see such behavioural changes as part of the instinctive make-up of human beings of whatever time period. The ability to change and adapt is one of the main elements which distinguished *homo sapiens* from the earlier hominids and from other animals – and has done since his evolution in the African heat and his move to, amongst other places, the ice of Europe. And, of course, there are also the two most visible distinguishing marks of mankind in which he is different from the rest – the utilisation of tools to increase his inborn capabilities, and his practice and appreciation of art, both of which elements are present from the very beginning of *homo sapiens'* existence (drawings on rock – using the stone as a tool – are known as early as 100,000 years ago in South African caves.)

Three fundamental elements are characteristic of mankind as revealed in the adaptations made in the Mesolithic period: the acceptance, if reluctantly, of change, the use of tools, and the appreciation and practice of art; to these may be added an active imaginative life, which manifested itself partly in art, partly in invention, partly in superstition and religion. These are, of course, still the determining elements of life in the Industrial Age. For we live in a time of incessant inventiveness and curiosity, increasing the number and range of the tools at our disposal, just as did the men of the Ice Age and after. Notice also that some of the inventions – radio, television, photography, printing – are purveyors of art in various forms to widely appreciative audiences. Television programmes and films are the industrial equivalent of painting on Ice Age cave walls, and they are a result of the same impulse towards artistic expression and the recording of the events of life.

So we may accept that the people who are facing the problems and uncertainties of the new episode of global warming have essentially the same capacities and capabilities as the people who had to cope with the vagaries of the post-Ice Age climate. The solutions to our new problem will, of course, have to be different from those developed in the Mesolithic period. We will not be developing a foraging way of life, do not have to (learn to) domesticate cattle or wheat or rice, or to invent farming. It may be that one result of the new warming will be to stimulate migrations into the new lands, which will become accessible in the northern and southern latitudes as the ground warms; this would be a modern version of the migration of the mammoth hunters northward, that is, it would be a refusal to change dressed up as the reclamation of new lands; it is also likely that there will be developed new strains of plants which can survive in the lands of increased heat – farmers have been breeding and cloning plants of various types for thousands of years.

The new solutions to the problem will have to grow out of the situation in which we are now, just as the hunters of the Ice Age had to start from where they were when they began to cope with the new climate. So it will be the industrial and technological inventiveness of humans which will be the way out of the problem, just as at the end of the Ice Age the invention of bows and arrows, and the use of microliths were technological responses to a new situation. We may also assume that most of the solutions which are tried, successful or not, will have the aim of changing people's lives as little as possible, even though we all live in a time when we have become unusually accepting of change, at least in relatively small things.

We have become adept at scientific and technological investigation and invention, at manufacturing, at economic manipulation, even at crisis management. Mankind is curious and ingenious. At the end of the Ice Age men invented a whole series of new ways of making a living in the process of adapting themselves to the new conditions, and the successful adaptations were the result of a long series of experiments. This acceptance of change is what people are good at, even if they grumble at the necessity, and that their aim is to conserve as much of their old lifestyles as possible. The methods of doing so in all cases were technological, by inventing new tools and practices, and new methods of getting food to assist in their survival – bows and arrows, pottery, agriculture, pastoralism, fishing techniques. This, like the well-developed art of the time, was an inheritance from the preceding Ice Age, when earlier needs had forced the inventions of other tools, such as microliths, and compelled men to paint pictures on cave walls. Again, this is what mankind is good at.

And this is what will happen in the future. It is ingenuity and inventiveness which will provide the means of survival and prosperity for people in the coming centuries, when, presumably, life on earth is lived in a warmer climate (though what goes up may come down – the present episode may be only temporary, and a new cold period may develop, just as temporary cold periods have happened in the past; one simply does not know – but note that the Bronze Age climate in Europe was probably warmer than now). For some, of course, this warming will be a welcome development, but for others less so. But in all cases technology will be applied to solve the problems. This is what men and women do. This is one of the 'lessons' of the earlier episode. Another is that there are always casualties in change, and mankind's great and unprecedented numbers have rendered it vulnerable – to hunger, to disasters, to new diseases – and it will be necessary at times to harden our hearts to suffering.

The problem will be over-confidence. Already there is hubristic talk among some scientists and engineers and journalists that the best solution will be to manipulate the climate of the earth as a whole, a 'solution' which will undoubtedly make matters worse, since any unintended, and unimagined, consequences – and there will be some – will soon distort the best of intentions. (Here, the basic conservative instincts of most people will no doubt act as a successful brake on the wilder notions.) A more certain approach will be to work slowly and gradually towards solutions which will cope with the immediate difficulties which arise, and so to avoid wholesale answers, and their inevitable mistakes. A variety of different solutions which were worked out in the post-Ice Age crisis – as in moving, foraging, cattle herding, fishing, farming – would be the best approach, since it is only by working out and testing such different answers that it can be discovered which of the suggested solutions is the best and most enduring. There will, of course, be many casualties along the way, like the Abu Hureyrans who were suffering from malnutrition at the end of their farming experiment, or the Australians who were isolated on the islands in the Bass Strait. The newly-rising sea level threatens numerous low-lying Pacific and Indian Ocean islands with complete destruction, but it also threatens to swamp many great cities, most of which are built deliberately to have access to the sea. Floods and extreme weather are likely to destroy resources, and famines are probable. Since one of the main causes of the problem is the great increase in the human population, disasters of various sorts will act as a partial corrective by killing many people. But a species which can justify the Great War and the use of atomic weapons can certainly take casualties – and the basic cause of the problem is the over-population of the Earth.

The adaptability and inventiveness which are so clear in the Mesolithic Age will be the prime requirements for surmounting the climate problems of the Industrial Age. The issue of global warming, in so far as it is a human product and not merely a natural alteration – or alternation – is one result of the industrialisation which has taken place over the last two centuries – just as is the understanding of what has occurred. This industrialisation has also been the main engine for lifting the great majority of humanity out of poverty, just as it has been the main source of the increase in numbers. So there can be no question of the Luddite response – destroy the machines – which some of the more extreme 'green' advocates suggest. (It would also be helpful if the gloom which many of them radiate could be controlled: why not some optimism that the problem can be addressed?) A Luddite destruction of industry might solve the global warming issue, though that is by no means certain. What is quite certain is that it will condemn the great majority of people in the world to an early death, and the majority of the survivors to poverty. It is necessary to start from where we are now, not turn back the clock to some mythical past and start again. So inventing new technological answers to the global problem will be the real overall solution, today as it was in the past.

The most successful answer to the post-Ice Age warming was eventually the invention of farming, which really meant two things: the successful recognition, isolation, and development of productive plants, and the creation of a technology to convert them to human use. The answers in the future will be similarly biological and technological. For example, since it is a global problem, one of the expedients is likely to be an exploration of the solar system and human settlement on the other planets, and perhaps space stations – the equivalent, in a sense, of the settlement of the new continent of America by the Beringians, or, earlier, the settlement of Australia 60,000 years ago. This will have the added advantage that if things go drastically wrong on the Earth human beings will survive elsewhere.

The only real 'lesson' which the global warming of the past has for us, therefore, is that experimentation and human adaptability are the keys to the present predicament, and that single solutions are not possible until the experimentation has taken place. We can be sure that the world after the warming will be different; but mankind will remain the same curious and artistic creature, if he survives.

Index

Abu Hureyra, Syria, 340, 238–41, 243, 247, 253, 263, 273, 283; figs 53, 55
Afghanistan, 232
Africa, 10, 12, 19, 24, 37, 43–4, 254–6; fig. 50
Ahrensburg culture, 66, 69, 153, 156
Alaska 34, 37, 38, 40, 44, 74, 88–93, 100, 108–109, 113–16, 120, 123, 127–8, 217; figs 28, 31
Aleutian Islands, 91, 100–101, 113, 116–17, 168
Aleuts, 91, 99, 117, 119, 123, 145, 167; fig. 23
Alexander the Great, 144, 148; fig. 36
Allahabad, India, 186
Alps Mtns, 38, 57
Altamira cave, Spain, 53, 61
Amur R., Siberia, 170, 176
Anatolia, 232, 236
Andes Mtns, 43, 141
Animal domestication, 216–21, 243–4, 245
Animal extinctions 126–7, 130–3, 202; Gallery IV
Antarctic, 10, 32, 44
Arabia, 10, 12, 37, 143, 148, 231
Arabian/Persian Gulf, 139–44, 162–5; fig. 35
Arabs, 147
Arafura Sea, 202
Arctic, 10
Arnhem land, Australia, 20, 23, 205, 209
Arrian, historian, 145, 147, 148
Artois, France, 58
Atlantic Ocean, 39, 105
Atlas Mtns, 215, 219
Australia, 4, 12–25, 36, 39, 43, 44, 47, 126, 134–6, 150, 157, 201–11, 221, 227; figs 3, 4, 48, 49; Gallery I

Babylonia, 144–5, 164, 229, 262; fig. 53
Badger I., Australia, 15–3
Bali, Indonesia, 14
Baltic Sea, 70, 101, 107, 154
Bass Strait, 22, 134, 150–63, 165, 204, 209; fig. 37
Bay of Biscay, 58
Bednarik, Professor Robert, 19
Beidha, Jordan 243–4
Beluga Point, Alaska, 121
Beringia, 44, 88–93, 100, 108, 111–13, 117, 123–4, 127–9, 136, 147; fig 22
Bering Strait, 22, 36, 38, 88, 114
Bir Kiseiba, Egypt, 216, 218
Black Sea, 38, 58
Border I., Australia, 210
Bosporos, 38
Brittany, France, 58
British Columbia, 101, 120–1, 123, 125
Bronze Age, 3, 31
Brooks Range, Alaska, 113, 117
Buka I., Melanesia, 22

Cactus Hill, Virginia, 111
Calais, France, 58
California, 122, 125–7, 193
Canada, 34, 37, 40, 88, 117, 128, 193
Carpathian Mtns, 58, 64
Carpenter's Gap, Australia, 24
Caspian Sea, 58
Caucasus Mtns, 58
cave dwelling 61
Cayonu, Turkey, 244
Charlie Lake, Canada, 117, 124
Central America, 254, 262
Chauvet cave, France, 52, 61
Chile, 109, 113, 119, 122, 196; fig. 47
China, 4, 24, 37–40, 44, 88, 176–7, 246–53; fig. 57

Chukotka, Siberia, 88–9, 108
Colombia, 113
Colonsay I., Scotland, 178, 181
Columbia R., U.S.A., 104
Cornwall, England, 58
Craig Point, Alaska, 121

Dalma, Abu Dhabi, 193
Dardanelles, 38
Denmark, 70, 74, 153–6, 158–63, 166,
    183–90, 194, 227; figs 38–40
Desna R., Ukraine 75; fig. 17
Diaotonghuan cave, China 246–53; figs 57–8
Doggerland, 36, 38, 58, 67–9, 71–3, 77,
    80, 84–9, 94, 106, 114, 123, 128, 138–9,
    149, 153–4, 183–9; 270–1, 273, fig. 16
Dolni Vestonice, Czech Rep., 62, 178
Don R., Ukraine, 74–5

East Africa, 211–12, 216–17; fig. 50
Egypt, 217, 262
Elbe R., Germany, 67
Ems R., Germany, 67
England, 57
English Channel, 38, 58, 67, 135
Europe, 4, 19, 23–5, 31, 34, 37–42, 44–5,
    47–8, 50, 57–8, 60–3, 68, 70–1, 73, 76–7,
    81, 84, 87–9, 94, 100, 106, 121, 126–9,
    156–7, 161, 168–9, 171, 186, 195, 202,
    210, 212–13, 227, 231, 261, 274–5,
    277–8, 280, 282; figs 10–11; Gallery II

Fezzan, Sahara, 215
Flinders Is., Australia, 150–3, 163
Fortification Point, New Guinea, 20
France, 57, 61
Fukui, Japan, 176

Gallagher Flint Station, Canada, 117
Galloway, Scotland, 182
Ganges R., India, 39, 186–7, 190,
    199–201, 277–8; fig. 46
Gedrosia, Iran, 144–50, 147–8, 153,
    159, 164, 166, 276; fig. 36
Germany, 57, 74
Global warming, causes, 1–5, 32
    effects, 4, 38–9
Gobekli, Turkey, 242–3, 267–8, 277
Great Bear Lake, Canada, 101

Great Lakes, North America, 101, 103, 105
Great Salt Lake, Utah, 103–104, 127
Great Slave Lake, Canada, 101
Greece, 37
Greenland, 32, 88, 100
Guangxi, China, 176
Gulf of Carpentaria, 134, 202
Gulf of St Lawrence, Canada, 137

Halmahera, Indonesia, 15; fig. 3
Hamburg, Germany, 64
Hebior, Wisconsin, 111; fig. 29
Hilo, Buka Is., Melanesia, 22; fig. 5
Himalaya Mtns, 39, 187
Honshu, Japan, 167, 170, 176
Hook Is., Australia, 210
Hopei, China, 176
Huai, R., China, 246, 278; fig. 57
Huanghe R., China, 251
Hudson's Bay, Canada, 101, 106–107
Hudson River, New York, 105
Human varieties, 10–12, 23, 31–2; fig. 2
Huon peninsula, New Guinea, 14–15, 20

Iberia, Mesolithic, Gallery III
Ice Age, 3–6, 9, 24, 31–56
    amelioration and changes,
        57–9, 73–4, 234–65
    human reactions, 76–93, 269–84; fig. 1
Ichthyophagoi, 145–50, 166; fig. 36
India, 10, 186–91, 201, 227;
    figs 45–6; Gallery V
Indian Ocean, 36, 39, 139, 144
Indonesia, 138
Indus River, 39, 145
Industrial Revolution, 2–3
Inuit, 90–1, 93, 95, 99, 101, 108, 117,
    123–4, 128, 147, 164, 272; fig. 23
Iran, 37, 144, 244; fig. 35
Ireland, 58
Italy, 38, 57

Japan, 4, 34, 60, 136–8, 167–78, 182–4,
    194–5, 227, 262; figs 33–4, 41–3
Jutland, Denmark, 153–4, 158

Kakadu, Australia, 26, 28–9, 205–207, 209
Kamchatka, Siberia, 60, 168
Keppel Is., Australia, 210

Kimberley, Australia, 24
King Is., Australia, 150–3, 163; fig. 37
Kinloch, Rhum, Scotland, 179
Kizojima, Japan, 167
Kodiak I, Alaska, 121
Korea, 38, 167–70
Ksar Atil, 231
Kuk, New Guinea, 256–61, 264; figs 60–1
Kuwait, 141, 143
Kyushu, Japan, 168, 174, 176

Las Caballos Cave, Spain, 96
Lake Agassiz, North America,
    103–106, 129; fig. 25
Lake Athabasca, Canada, 101
Lake Bonneville, Utah, 103–106, 129
Lake Chad, Africa, 39, 212
Lake Eyre, Australia, 210
Lake Flixton, England, 73
Lake Missoula, North America,
    103–106, 108, 129; fig. 26
Lake Mungo, Australia, 14,
    18–20, 23, 27, 194, 210
Lake Naivasha, East Africa, 212
Lake Nakuru, East Africa, 212
Lake Ojibway, North America, 103, 105
Lake Rudolf, East Africa, 212
Lake Winnipeg, Canada, 101, 104
Laki, Iceland, eruption, 34, 36, 51; fig 6
Lascaux cave, France, 54–6, 61
Levant, 229–46, 262
Lombok, Indonesia, 10

Maine, 137
Malaysia, 10, 14, 138
Malakunanja, Australia, 20, 23; figs 4, 49
Makenkuptum, Melanesia, 22
Manchuria, China, 176
Manus Is., Melanesia, 22; fig. 5
Meadowcroft Shelter, Pennsylvania,
    109, 111; fig. 29
Mediterranean Sea, 38, 216
Meiendorf, Germany, 64, 67, 70–1,
    75, 88, 114, 153, 184; figs 15–16
Melanesia, 20–2, 24–5, 44; fig. 5
Melbourne, Australia, 17
Merewah Is., Abu Dhabi, 141–3; fig. 35
Mesolithic, 5–7, 48, 87, 157; Gallery III
Mesopotamia, 231

Mexico, 113, 125, 262
Miasma theory, 12
Middle East, 4
Migrations, 43, 216–17
Mississippi R., 109
Moluccas, Indonesia, 14
Mongolia, 100
Monte Bello Is., Australia, 208–10
Monte Verde, Chile, 109, 113, 129; fig. 30
Moravia, 62
Mostin, California, 122
Mount Carmel, Israel, 231
Myrvatnet, Norway, 80; fig. 18

Nabta, Egypt, 216, 218; fig. 52
Narbada R., India, 190
Natufians, 237–8, 240–3, 246,
    262–6, 271; figs 56, 61
Nearchos, historian, 145, 147, 148
Nenana River, Alaska, 113–16, 125; fig 28
New Britain, Melanesia, 16, 20, 22; figs 3, 5
Newfoundland, 136
New Guinea, 12, 14–15, 20, 24, 38, 44,
    134, 138, 201, 204, 256–61; figs, 3–5, 60
New Ireland, Melanesia, 22
New South Wales, 14–15, 211
Neolithic, 5, 156, 245
Niaux cave, Belgium, 55
Nile R., 39, 218
North Africa, 39, 232
North America, 43 60, 101, 134, 157
    peopling of, 108–29, 227
    ice-free corridor, 109–113; figs 13, 27, 32
North Sea, 36, 38, 58, 67, 71, 73,
    107, 135, 153–4, 161, 163
Norway, 77–87, 107, 123, 128, 139,
    156, 184; figs 18–19, 21
Nova Scotia, 137
Novgorod, Russia, 62–3

Obsidian, 22, 48, 121, 167, 169–70
Ohalo, 234–8, 265; figs 53–4
Oman, 139, 141, 143
Oregon, 125
Oronsay Is., Scotland, 59, 161,
    178–86, 237, 275–6; fig. 44

Pacific Ocean, 22, 39, 101, 168, 191
Palaeolithic, 4–6, 19, 22, 25, 48, 51, 53,
    66, 94, 147, 156–7, 167, 173–4, 176–8,
    182, 184–90, 201, 250, 270–1, 274–5

Palestine, 227
Persian Gulf, 36–7
Peru, 191–3, 208, 262, 269, 275; fig. 47
Philippines, 12, 25, 38
Plant domestication, 227–63; Gallery VII
Polynesia, 22
Population growth, 3
   spread, 10–12, 15–25, 37; fig. 2
Prime Seal Is., Australia, 151–3
Puget Sound, 125
Pushkari, Russia, 62–8; figs 14, 17
Pyrenees Mtns, 38, 57

Queensland, 14, 204–205, 210

Rhine R., 67
Rhum Is., Scotland, 179–81
Rocky Mtns, 88, 99, 101, 103,
   116–17, 120–1, 125
Russia, 37, 57–8, 60, 73
   Far East, 60, 88, 100, 176–7; fig. 17

Sahara, 4, 39, 211–21, 227, 253–6;
   figs 50–2; Gallery VI
Sahel, Africa, 211, 220
Sahul, 12, 14–15, 20, 36, 43, 211; figs 3, 12
St Lawrence River, North America, 105
Sakhalin, 168
Saltville, Virginia, 111
San Francisco Bay, 124–5
Santa Rosa Is., California, 122
Scandinavia, 38, 57–8, 60, 67, 70–1,
   73–4, 77, 80, 83–4, 99, 107, 128,
   134, 158, 186, 269; fig. 16
Schaefer, Wisconsin, 112; fig. 29
Scotland, 57, 59, 78, 107, 134, 139, 178
sea level changes, 59, 167, 202–204
Sea of Japan, 36, 38, 167
Seward Peninsula, Alaska, 117
Shanxi, China, 176
Shetland Is., Scotland, 38
Shikoku, Japan, 168, 176
Siberia, 34, 37–8, 40, 44, 60, 61,
   73–4, 88, 99, 127–8, 169–70
Sinai, 217–18, 221, 243, 277
Slettnes, Norway, 80; figs 18–19
South Africa, 49

South America, 43, 50, 139, 194, 228
   peopling of, 109
Spain, 38
Star Carr, England, 66, 69,
   71, 81, 156–7, 190
Stellmoor, Germany, 64, 66–7, 70,
   75, 153, 158, 184; fig. 15
Storegga Slide, Norway, 107
Strabo, historian, 145–6
Strait of Tsushima, 168–9
Sudan, 216, 218
Sulawesi, Indonesia, 10, 12,
   14–15, 17, 138; fig. 3
Sundaland, 10, 12, 14, 36, 38,
   138, 150, 202; fig. 3
Sweden, 77, 107
Sydney, Australia, 17
Syria, 231

Taiwan, 38
Tanana River, Alaska, 113–14, 125
Tasmania, 12, 14, 24, 38, 43,
   134, 150, 152, 181, 204
Tassili Mtns, 215
Temperature variation, 3–4, 9
Thailand, 138
Thames River, 67
Tierra del Fuego, 193
Toba, Sumatra, eruption, 34, 51

Ukraine, 74–6, 89
Umm al-Qaiwan, 141–2; fig. 35
Umm Qatafa, 231
United States, 43, 110–11, 121, 125–6, 129

Vermilion Lake, Canada, 117, 120, 124
Vindya Hills, India, 186, 188

Weald, England, 58
Western Australia, 208–209
Willandra Lakes, Australia, 26–7, 210

Yangzi R., China, 246–7, 249, 278; fig. 57
Yoldia Sea, 107
Yombon, 22; fig. 5

Zuider Zee, 67